Lecture Notes in Computer Science 6865

Commenced Publication in 1973
Founding and Former Series Editors:
Gerhard Goos, Juris Hartmanis, and Jan van Leeuwen

Christian Böhm   Sami Khuri
Lenka Lhotská   Nadia Pisanti (Eds.)

Information Technology in Bio- and Medical Informatics

Second International Conference, ITBAM 2011
Toulouse, France, August 31 - September 1, 2011
Proceedings

 Springer

Volume Editors

Christian Böhm
Ludwig-Maximilians-Universität, Department of Computer Science
Oettingenstrasse 67
80538 München, Germany
E-mail: boehm@dbs.ifi.lmu.de

Sami Khuri
Department of Computer Science, San José State University
One Washington Square
San José, CA 95192-0249, USA
E-mail: khuri@cs.sjsu.edu

Lenka Lhotská
Czech Technical University
Faculty of Electrical Engineering, Department of Cybernetics
Technicka 2
166 27 Prague 6, Czech Republic
E-mail: lhotska@fel.cvut.cz

Nadia Pisanti
Dipartimento di Informatica, Università di Pisa
Largo Pontecorvo 3
56127 Pisa, Italy
E-mail: pisanti@di.unipi.it

ISSN 0302-9743   e-ISSN 1611-3349
ISBN 978-3-642-23207-7   e-ISBN 978-3-642-23208-4
DOI 10.1007/978-3-642-23208-4
Springer Heidelberg Dordrecht London New York

Library of Congress Control Number: 2011933993

CR Subject Classification (1998): H.3, H.2.8, H.4-5, J.3

LNCS Sublibrary: SL 3 – Information Systems and Application, incl. Internet/Web
and HCI

Typesetting: Camera-ready by author, data conversion by Scientific Publishing Services, Chennai, India

Printed on acid-free paper

Springer is part of Springer Science+Business Media (www.springer.com)

Preface

Biomedical engineering and medical informatics represent challenging and rapidly growing areas. Applications of information technology in these areas are of paramount importance. Building on the success of the first ITBAM that was held in 2010, the aim of the second ITBAM conference was to continue bringing together scientists, researchers and practitioners from different disciplines, namely, from mathematics, computer science, bioinformatics, biomedical engineering, medicine, biology, and different fields of life sciences, so they can present and discuss their research results in bioinformatics and medical informatics. We trust that ITBAM served as a platform for fruitful discussions between all attendees, where participants could exchange their recent results, identify future directions and challenges, initiate possible collaborative research and develop common languages for solving problems in the realm of biomedical engineering, bioinformatics and medical informatics. The importance of computer-aided diagnosis and therapy continues to draw attention worldwide and has laid the foundations for modern medicine with excellent potential for promising applications in a variety of fields, such as telemedicine, Web-based healthcare, analysis of genetic information and personalized medicine.

Following a thorough peer-review process, we selected 13 long papers and 5 short papers for the second annual ITBAM conference. The Organizing Committee would like to thank the reviewers for their excellent job. The articles can be found in these proceedings and are divided into the following sections: decision support and data management in biomedicine; medical data mining and information retrieval; workflow management and decision support in medicine; classification in bioinformatics; data mining in bioinformatics. The papers show how broad the spectrum of topics in applications of information technology to biomedical engineering and medical informatics is.

The editors would like to thank all the participants for their high-quality contributions and Springer for publishing the proceedings of this conference. Once again, our special thanks go to Gabriela Wagner for her hard work on various aspects of this event.

June 2011

Christian Böhm
Sami Khuri
Lenka Lhotská
Nadia Pisanti

Organization

General Chair

Christian Böhm   University of Munich, Germany

Program Chairs

Sami Khuri   San José State University, USA
Lenka Lhotská   Czech Technical University Prague,
         Czech Republic
Nadia Pisanti   University of Pisa, Italy

Poster Session Chairs

Vaclav Chudacek   Czech Technical University in Prague,
         Czech Republic
Roland Wagner   University of Linz, Austria

Program Committee

Werner Aigner   FAW, Austria
Fuat Akal   Functional Genomics Center Zurich,
         Switzerland
Tatsuya Akutsu   Kyoto University, Japan
Andreas Albrecht   Queen's University Belfast, UK
Julien Allali   LABRI, University of Bordeaux 1, France
Lijo Anto   University of Kerala, India
Rubén Armañanzas Arnedillo   Technical University of Madrid, Spain
Peter Baumann   Jacobs University Bremen, Germany
Balaram Bhattacharyya   Visva-Bharati University, India
Christian Blaschke   Bioalma Madrid, Spain
Veselka Boeva   Technical University of Plovdiv, Bulgaria
Gianluca Bontempi   Université Libre de Bruxelles, Belgium
Roberta Bosotti   Nerviano Medical Science s.r.l., Italy
Rita Casadio   University of Bologna, Italy
Sònia Casillas   Universitat Autònoma de Barcelona, Spain
Kun-Mao Chao   National Taiwan University, China
Vaclav Chudacek   Czech Technical University in Prague,
         Czech Republic

Table of Contents

Decision Support and Data Management in Biomedicine

Medical Data Mining and Information Retrieval

Workflow Management and Decision Support in Medicine

Classification in Bioinformatics

Data Mining in Bioinformatics

Exploitation of Translational Bioinformatics for Decision-Making on Cancer Treatments

Jose Antonio Miñarro-Giménez[1], Teddy Miranda-Mena[2],
Rodrigo Martínez-Béjar[1], and Jesualdo Tomás Fernández-Breis[1]

[1] Facultad de Informática, Universidad de Murcia, 30100 Murcia, Spain
{jose.minyarro,rodrigo,jfernand}@um.es
[2] IMET, Paseo Fotografo Verdu 11, 30002 Murcia, Spain
teddygonzalo.miranda@um.es

Abstract. The biological information involved in hereditary cancer and medical diagnoses have been rocketed in recent years due to new sequencing techniques. Connecting orthology information to the genes that cause genetic diseases, such as hereditary cancers, may produce fruitful results in translational bioinformatics thanks to the integration of biological and clinical data. Clusters of orthologous genes are sets of genes from different species that can be traced to a common ancestor, so they share biological information and therefore, they might have similar biomedical meaning and function.

Linking such information to medical decision support systems would permit physicians to access relevant genetic information, which is becoming of paramount importance for medical treatments and research. Thus, we present the integration of a commercial system for decision-making based on cancer treatment guidelines, ONCOdata, and a semantic repository about orthology and genetic diseases, OGO. The integration of both systems has allowed the medical users of ONCOdata to make more informed decisions.

Keywords: Ontology, Translational bioinformatics, Cluster of Orthologs, Genetic Diseases.

1 Introduction

Translational bioinformatics is involved in the relation of bioinformatics and clinical medicine. Bioinformatics was originated by the outstanding development of information technologies and genetic engineering, and the effort and investments during the last decades have created strong links between Information Technology and Life Sciences Information technologies are mainly focused on routine and time-consuming tasks that can be automated. Such tasks are often related to data integration, repository management, automation of experiments and the assembling of contiguous sequences. On the medical side, decision support systems for the diagnosis and treatment of cancers are an increasingly important factor for the improvement of medical practice [1][2][3][4]. The large amount of

C. Böhm et al. (Eds.): ITBAM 2011, LNCS 6865, pp. 1–15, 2011.

information and the dynamic nature of medical knowledge involve a considerable effort to keep doctors abreast of medical treatments and the latest research on genetic diseases. These systems have proven to be beneficial for patient safety by preventing medication errors, improving health care quality through its alignment with clinical protocols and making decisions based on evidences, and by reducing time and costs.

In modern biomedical approaches, bioinformatics is an integral part of the research of diseases [5]. These approaches are driven by new computational techniques that have been incorporated for providing general knowledge of the functional, networking and evolutionary properties of diseases and for identifying the genes associated with specific diseases.

Moreover, the development of large-scale sequencing of individual human genomes and the availability of new techniques for probing thousands of genes provide new biological information sources which other disciplines, such medicine, may and even need to exploit. Consequently, a close collaboration between bioinformatics and medical informatics researchers is of paramount importance and can contribute to a more efficient and effective use of genomic data to advance clinical care [6].

Biomedical research will also be powered by our ability to efficiently integrate and manage the large amount of existing and continuously generated biomedical data. However, one of the most relevant obstacles in translational bioinformatics field is the lack of uniformly structured data across related biomedical domains [7]. To overcome this handicap, the Semantic Web [8] provides standards that enable navigation and meaningful use of bioinformatics resources by automatic process. Thus, translational bioinformatics research, with the aim to integrate biology and medical information and to bridge the gap between clinical care and medical research, provides a large and interesting field for biomedical informatics researchers [9].

The research work described in this paper extends a commercial decision-making system on cancer treatments, the ONCOdata system [10], which has been used in the last years in a number of oncological units in Spain. In silico studies of the relationships between human variations and their effect on diseases have be considered key to the development of clinically relevant therapeutic strategies [11]. Therefore, including information of the genetic component of the diseases addressed by the professionals who are using ONCOdata was considered crucial for adapting the system to state-of-the-art biomedical challenges.

To this end we have used the OGO system [12], which provides integration information on clusters of orthologous genes and the relations between genes and genetic diseases. Thus, we had to develop methods for the exchange of information between two heterogeneous systems. OGO is based on semantic technologies whereas ONCOdata was developed using more traditional software technologies, although it makes use of some expert knowledge in the form of rules and guidelines.

The structure of the rest of the paper is described next. First, the background knowledge and the description of the systems used for this translational experience are presented in Section 2. Then, the method used for the exchange of information in Section 3, whereas the results will be presented in Section 4. Some discussion will be provided in Section 5. Finally, the conclusions will be put forward in Section 6.

2 Background

The core of this research project comprises the two systems that will be interconnected after this effort. On the one hand, ONCOdata is a commercial system that supports medical doctors on decision-making about cancer treatments. Thus, it is an intelligent system which facilitates decision-making based on medical practice and medical guidelines. On the other hand, OGO provides an integrated knowledge base about orthology and hereditary genetic diseases. OGO uses Semantic Web Technologies for representing the biomedical knowledge for integrating, managing and exploiting biomedical repositories.

The next subsections go through some of the functionalities of the systems and provide the technical details that differentiate both systems. The first subsection describes the different modules of ONCOdata, whereas the second subsection provides a brief description of the OGO system.

2.1 The ONCOdata System

The ONCOdata application is a decision support system which helps to allocate cancer treatments via Internet. In particular, ONCOdata is divided into two main modules, namely, ONCOdata record and ONCOdata decision.

ONCOdata record implements the management information of cancer health records. This module is responsible for storing the information produced in all cancer stages, beginning with the first medical visit and continuing with diagnosis, treatment and monitoring. The information produced in each stage is suitable managed and organized in the cancer health record of ONCOdata record module. This system does not use any electronic healthcare records standards like HL7[1] , openEHR[2] or ISO 13606[3] but a proprietary one. Fortunately, this module is able to generate standardized contents using the MIURAS integration engine[13].

On the other hand, ONCOdata decision is responsible for supporting physicians in making appropriate decisions on cancer treatments. It provides details of patient cancer subtype, so the physician may make informed decisions of which treatment should be applied in each case. In this way, the module recommends the best treatments based on the patient's cancer health record. For this purpose, ONCOdata uses the representation of each patient's cancer disease based on their medical and pathological information. Then, its reasoning engine uses this representation and a set of expert rules to generate the recommendations.

---

[1] http://www.hl7.org
[2] http://www.openehr.org
[3] http://www.en13606.org

The knowledge base used by the reasoning engine was developed by a group of cancer domain experts and knowledge management experts, which acted as consultants for the company. The knowledge base was technically built by using Multiple Classification Ripple Down Rules[14]. Besides, the development and maintenance of the knowledge base follows an iterative and incremental process.

Physicians use ONCOdata through a web interface that allows them to insert the patient's medical information, and then to retrieve the recommendations about the suitable treatment. Not only recommendations are provided, but also the evidences and bibliographic materials that support those recommendations. Thus, physicians, after gathering information from cancer patients, can find medical advice from the ONCOdata system which facilitates making the decision on cancer treatment. This process is described in Figure 1.

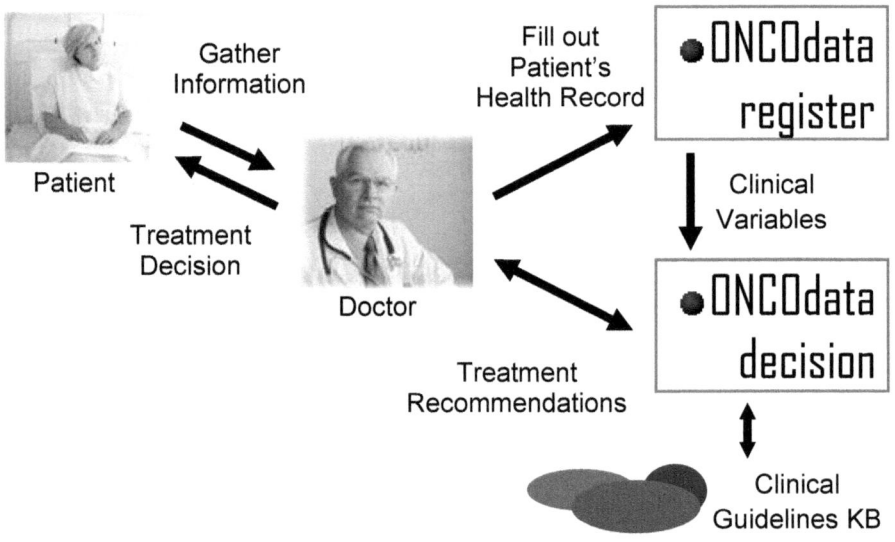

Fig. 1. The ONCOdata system

ONCOdata was designed to be useful for doctors in every disease stage. For example, during the breast cancer process a multiskilled team of cancer experts, each responsible for a different medical area, is involved in the treatment of patients. Figure 2 shows the various medical areas that are involved in the breast cancer process treatment. ONCOdata provides clinical records to store and manage the information produced during every disease stage and therefore, the opportunity to use such information for making treatment decisions.

2.2 The OGO System

The Ontological Gene Orthology (OGO) system was first described in [15]. This system was initially developed for integrating only biological information about

Fig. 2. The Breast Cancer Process

orthology. Then, information sources about genetic diseases were also integrated to covert it in a translational resource. The OGO system provides information about orthologous clusters, gene symbols and identifiers, their organism names and their protein identifiers and accession numbers, genetic disorders names, the genes involved in the diseases, their chromosome locations and their related scientific papers.

The information contained in the OGO system is retrieved from the following publicly available resources: KOGs[4], Inparanoid[5], Homologene[6], OrthoMCL[7] and OMIM[8]. The first four resources contain information about orthology, whereas OMIM provides a continuously updated authoritative catalogue of human genetic disorders and related genes. Therefore, the development of the OGO system demanded the definition of a methodology for integrating biological and medical information into a semantic repository, which is described in [12].

The design, management and exploitation of the OGO system is based on Semantic Web Technologies. Thus, a global ontology (see Figure 3) becomes the cornerstone of the OGO system and which reuses other bio-ontologies, such as the Gene Ontology (GO)[9], the Evidence Code Ontology (ECO)[10] and the NCBI species taxonomy[11]. This global ontology defines the domain knowledge of

---

[4] http://www.ncbi.nlm.nih.gov/COG/
[5] http://inparanoid.sbc.su.se/cgi-bin/index.cgi
[6] http://www.ncbi.nlm.nih.gov/homologene
[7] http://orthomcl.org
[8] http://www.ncbi.nlm.nih.gov/omim/
[9] http://www.geneontology.org/
[10] http://www.obofoundry.org/cgi-bin/detail.cgi?id=evidence_code
[11] http://purl.bioontology.org/ontology/NCBITaxon

orthologous genes and genetic diseases. This ontological knowledge base is then populated through the execution of the data integration process. The proper semantic integration is basically guided by the global ontology. The definition of the OGO ontology also includes restrictions to avoid inconsistencies in the OGO KB. The restrictions defined in the ontology were basically disjointness, existential qualifiers (to avoid inconsistencies in the range of object properties); and cardinality constraints. The Jena Semantic Web Framework[12] is capable of detecting such issues, therefore its usage facilitates checking the consistency of the ontology when used together with reasoners, such as Pellet[13]. The OGO KB contains more than 90,000 orthologous clusters, more than a million of genes, and circa a million of proteins. Besides, from the genetic diseases perspective it contains approximately 16,000 human genetic disorders instances and more than 17,000 references to scientific papers.

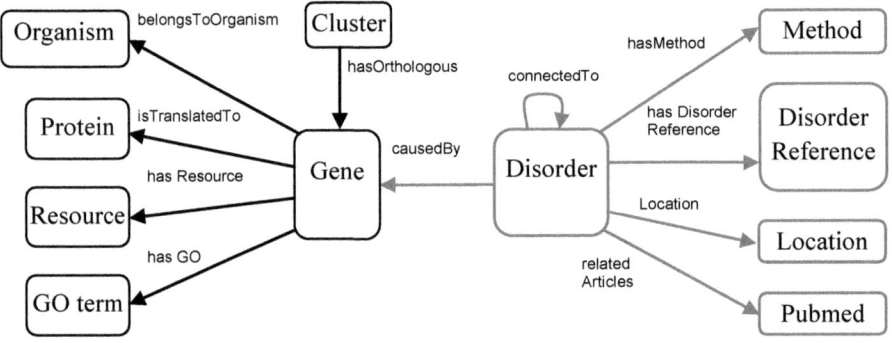

Fig. 3. The OGO ontology

The users of the OGO system can navigate through the genes involved in a particular genetic disorder to their orthologous clusters and vice versa using the ontology relations and concepts. The web interfaces developed for querying OGO KB allow data exploitation from two complementary and compatible perspectives: orthology and genetic diseases. For a particular gene, not only the information about its orthologous genes can be retrieved, but also its related genetic disorders. The search functionality for diseases is similar. The interfaces are based on web technology that allows non-expert users to define their query details[16]. Then, SPARQL queries are defined at runtime by the application server and hence the information is retrieved from the semantic repository. The more sophisticated the query is, the more exploitable the OGO KB is, so we have also developed a query interface for allowing more advanced SPARQL query definitions. The interface is driven by the OGO ontology during the query definition, providing users with all possible query options at each definition step.

---

[12] http://jena.sourceforge.net
[13] http://www.mindswap.org/2003/pellet/

3   Information Exchange between ONCOdata and OGO

In this section we describe the scope of the exchange information between the system and the details of how the communication process has been developed. First, we describe the approach followed in this work for establishing the communication between both systems. Second, we describe how the OGO system makes available its KB to external applications. Third, we describe how the ONCOdata system exploits the OGO KB functionalities. Finally, we describe the technical details of the communication module and its query interfaces and evaluate the results.

3.1   The Approach

As it has been aforementioned, ONCOdata and OGO are two completely separate applications, thus a solution with minimum coupling between the systems was required. Several technologies for interoperability between applications, such as XML-RPC[14], RMI[15], CORBA[16] or Web Services[17], were evaluated. This evaluation pointed out that the features of web services are the most suitable for the project requirements. Web services provide loosely coupled communication, and text-encoded data and messages. The widespread adoption of SOAP[18] and WSDL[19] standards together with HTTP[20] and XML[21] facilitate developers to adopt and less costly to deploy web services.

From a technical point of view, WSDL defines an XML grammar for describing network services as collections of communication endpoints capable of exchanging messages. On the other hand, SOAP describes data formats, and the rules for generating, exchanging, and processing such messages. Finally, HTTP was the chosen transport protocol for exchanging SOAP messages.

The system scenario is depicted in Figure 4. There, web services allow applications to query the information available in OGO. The OGO system then would process the query and define the SPARQL queries for providing the demanded information. Then, OGO returns to ONCOdata the client a XML document with that information. This solution has been developed for and applied to the exchange of information between ONCOdata and OGO although both systems would be able to exchange information with other systems by reusing the approach and the already available communication mechanisms.

3.2   Usage of OGO from other Applications

A series of web services have been developed to facilitate applications to query the OGO knowledge base. In particular, three web services have been developed

---

[14] http://www.xmlrpc.com/
[15] http://download.oracle.com/javase/tutorial/rmi/overview.html
[16] http://www.corba.org/
[17] http://www.w3.org/TR/2007/REC-ws-policy-20070904/
[18] http://www.w3.org/TR/2007/REC-soap12-part0-20070427/
[19] http://www.w3.org/TR/2007/REC-wsdl20-primer-20070626/
[20] http://www.w3.org/Protocols/
[21] http://www.w3.org/XML/

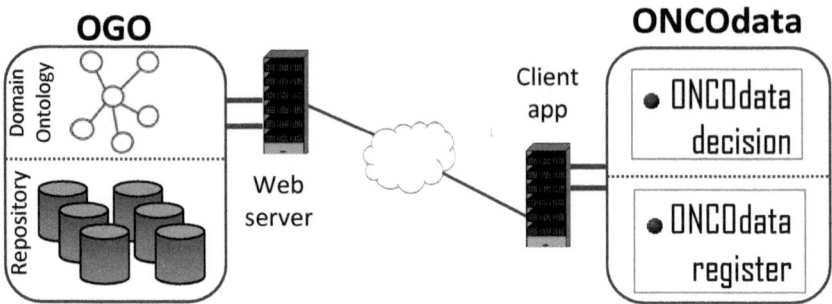

Fig. 4. The Integration Scenario

to achieve this goal: (1) service for querying orthology information by using gene names and its corresponding organism; (2) service for querying information about genetic diseases by using disease names; and (3) service for querying the OGO knowledge base by using user-defined SPARQL queries. OGO sends the results in XML documents, whose structure depends on the service that was invoked:

- Orthology information: the returned document will consist of all the genes of the same cluster of orthologous genes to which the input gene belongs, together with their relationships and information about properties.
- Genetic disease information: the returned document will contain information about the properties and relations of all diseases whose names match the disease names provided by the client application.
- SPARQL queries: the returned document contains the bindings for the variables defined in each query.

3.3   Integration in ONCOdata

Before ONCOdata suggests a suitable treatment for a patient, the case study must be inserted by using the ONCOdata record module. Each record consists of several clinical and pathological variables. These variables will be used by the reasoning engine to recommend one or more treatments. The knowledge of cancer treatments was extracted from clinical guidelines that doctors use to manually seek for clinical treatments. Thus, the OGO system can provide additional knowledge by linking information about diseases and genes. Consequently, the physicians may make more informed decisions.

The integration of the information from the OGO KB offers detailed information of genes and mutation locations related to hereditary diseases. During the early stages of the disease diagnosis, the physicians collect information about the familiar clinical record of the patients. Then, during the late stages of diagnosis, they complete the information of the health record of patients. After completing the health record and before making the decision of the treatment, having the genetic information related to the disease is prominent. Thus, doctors may be more supported to choose when making their decisions.

Figure 5 depicts one particular scenario of the exchange of information between OGO and ONCOdata about breast cancer. In this case, the physician, upon the completion of the patient's clinical record, uses the ONCOdata decision web interface for retrieving the suitable treatment recommendations for the patient. First, ONCOdata decision retrieves the case study from the patient's medical history from ONCOdata record. If the case study contains any hereditary risk of cancer, ONCOdata decision seeks for the breast cancer disease information from the OGO system. As a result of this service invocation, the information of the different diseases and the corresponding genes is retrieved. Next, ONCOdata infers and shows the treatment recommendations as well as the biomedical information associated with the disease. Finally, the physician selects a treatment, which is then recorded in the patient's clinical record using ONCOdata record.

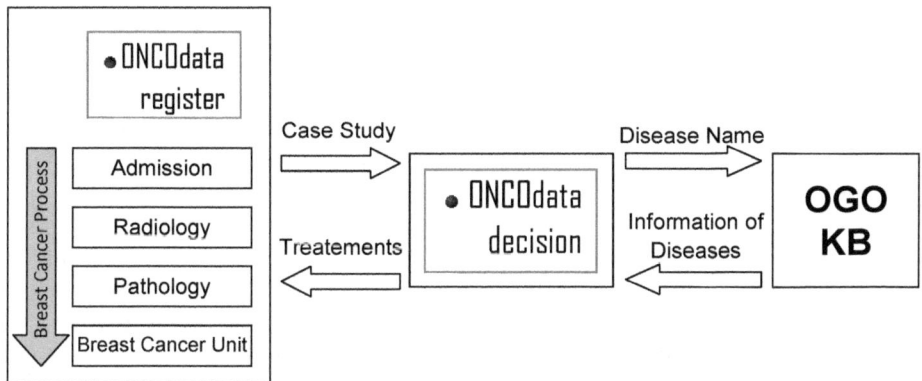

Fig. 5. The ONCOdata module

4 Results

As mentioned, ONCOdata can query the OGO knowledge base by invoking the web services developed for retrieving information on genetic diseases and orthologous genes. Figure 6 represents the different web service implemented for querying OGO. Thus, the *getDiseaseInformation* method interface is responsible for retrieving information about genetic diseases, the *getOrthologsInformation* method retrieves information about cluster of orthologous genes, and finally the *getSPARQLInformation* method which allow client to define their own SPARQL queries. In addition, if web service clients combine the first and second method interfaces, they can retrieve translational information from both perspectives, genetic disease and orthologs. However, if clients use the third method, they can obtain the specific information they want in one single query.

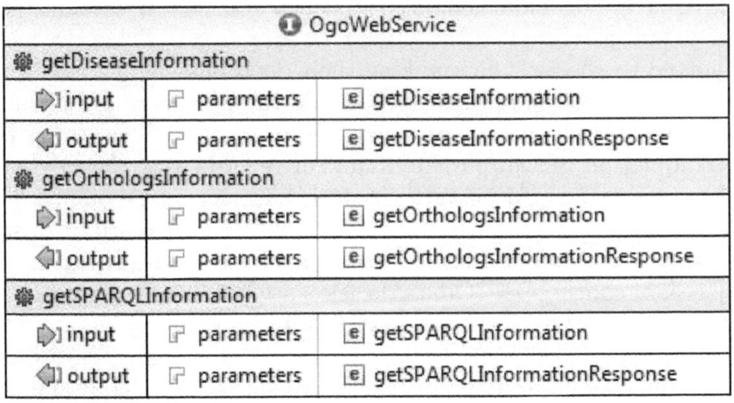

🔵 OgoWebService		
⚙ getDiseaseInformation		
▷] input	⬚ parameters	ⓔ getDiseaseInformation
◁] output	⬚ parameters	ⓔ getDiseaseInformationResponse
⚙ getOrthologsInformation		
▷] input	⬚ parameters	ⓔ getOrthologsInformation
◁] output	⬚ parameters	ⓔ getOrthologsInformationResponse
⚙ getSPARQLInformation		
▷] input	⬚ parameters	ⓔ getSPARQLInformation
◁] output	⬚ parameters	ⓔ getSPARQLInformationResponse

Fig. 6. The deployed web services

The web services are described using WSDL documents. In particular, the *getDiseaseInformation* web service is described by the WSDL document shown in Figure 7. We can see in this figure that this WSDL document defines the location of the service, namely, `http://miuras.inf.um.es:9080/OgoWS/services/OgoDisease`. In other words, a client can invoke this web service using the communication protocol description of such WSDL document. Therefore, clients and server applications may compose the proper request and response SOAP messages to communicate.

These services consist of one request and one response messages which are exchanged between the client and the server. In particular, when seeking for genetic disease information, ONCOdata sends the request message with the genetic disease name to OGO. Then, OGO, using the client query parameters with the genetic disease name and a pre-defined SPARQL query pattern (see Figure 8), generate the final SPARQL query. The query is defined and executed to retrieve all diseases instance, their related relationships and properties from the OGO knowledge base. Once the server obtains the query results, the data of each disease is encoded in the XML document, which is then sent back to ONCOdata.

Let us consider now the SPARQL query-based service. In this modality, the variables used in the query pattern, which is shown in Figure 8, represent the relationships and properties related to the disease class of the OGO ontology. Thus, the element nodes of the XML document are the data of the query variable. The root nodes of the XML document correspond to disease instances, and their child nodes correspond to their relationships and properties. Figure 9 shows an excerpt of the returned XML document which is generated when seeking for information on breast cancer. Finally, the XML document is processed by ONCOdata and displayed to the user.

The resulting system has been validated by the medical consultant of the company. For this purpose, a series of tests were designed by them and were systematically executed. They did validate that the new information was really useful from the medical perspective to support clinical practice.

```
<?xml version="1.0" encoding="UTF-8"?>
<wsdl:definitions targetNamespace="http://model" xmlns:apachesoap="http://xml.apache.org/xml-soap"
 xmlns:impl="http://model" xmlns:intf="http://model" xmlns:wsdl="http://schemas.xmlsoap.org/wsdl/"
 xmlns:wsdlsoap="http://schemas.xmlsoap.org/wsdl/soap/" xmlns:xsd="http://www.w3.org/2001/XMLSchema">
 <wsdl:types>
  <schema elementFormDefault="qualified" targetNamespace="http://model" xmlns="http://www.w3.org/2001/XMLSchema">
   <element name="getDiseaseInformation">
    <complexType><sequence><element name="diseaseName" type="xsd:string"/></sequence></complexType>
   </element>
   <element name="getDiseaseInformationResponse">
    <complexType><sequence><element name="getDiseaseInformationReturn" type="xsd:string"/></sequence></complexType>
   </element>
 </wsdl:types>
 <wsdl:message name="getDiseaseInformationRequest">
 <wsdl:message name="getDiseaseInformationResponse">
 <wsdl:portType name="OgoWebService">
  <wsdl:operation name="getDiseaseInformation">
 </wsdl:portType>
 <wsdl:binding name="OgoWebServiceSoapBinding" type="impl:OgoWebService">
  <wsdlsoap:binding style="document" transport="http://schemas.xmlsoap.org/soap/http"/>
  <wsdl:operation name="getDiseaseInformation">
   <wsdlsoap:operation soapAction=""/>
   <wsdl:input name="getDiseaseInformationRequest">
   <wsdl:output name="getDiseaseInformationResponse">
  </wsdl:operation>
 </wsdl:binding>
 <wsdl:service name="OgoWebServiceService">
  <wsdl:port binding="impl:OgoWebServiceSoapBinding" name="OgoWebService">
   <wsdlsoap:address location="http://miuras.inf.um.es:9080/OgoWS/services/OgoWebService"/>
  </wsdl:port>
 </wsdl:service>
</wsdl:definitions>
```

Fig. 7. The WSDL document for querying on genetic diseases

```
PREFIX ogo:<http://miuras.inf.um.es/ontologies/OGO.owl#>
PREFIX rdf:<http://www.w3.org/1999/02/22-rdf-syntax-ns#>
PREFIX rdfs:<http://www.w3.org/2000/01/rdf-schema#>
SELECT ?disease ?name ?location ?gene ?method ?article
WHERE {
   ?disease rdfs:label ?name .
   FILTER   regex(?name, "disease name", i) .
   ?disease ogo:location ?location .
   ?disease ogo:causedBy ?gene .
   ?disease ogo:hasMethod ?method .
   ?disease ogo:relatedArticles ?article .
   ?disease rdf:type ogo:Disorder .
   ?gene    rdf:type ogo:Gene .
   ?article rdf:type ogo:PubMed .
}
```

Fig. 8. The SPARQL query pattern used for querying on genetic disease information

```
<?xml version='1.0' encoding='iso-8859-1'?>
<reply>
  <disease>
    <geneName>SH2D3B</geneName>
    <geneName>NSP2</geneName>
    <geneName>SH2D3B</geneName>
    <location>1p22.1</location>
    <name>breast cancer anti-estrogen resistance 3</name>
    <pubmed>9582273</pubmed>
  </disease>
  <disease>
    <geneName>DKFZP564A063</geneName>
    <geneName>brms1</geneName>
    <location>11q13.1-q13.2</location>
    <name>Breast cancer metastasis suppressor 1</name>
    <method>In situ DNA-RNA or DNA-DNA annealing</method>
    <pubmed>10850410</pubmed>
  </disease>
</reply>
```

Fig. 9. Excerpt of the XML document returned when seeking for breast cancer disease

5   Discussion

Decision support systems play an increasingly important role to assign medical treatments to patients. Such systems increase the safety of patients by preventing medical errors, and facilitate decision-making processes by reducing the time in seeking for the most appropriate medical treatment.

In this way, ONCOdata is a decision-making system for the allocation of cancer treatments based on evidences. The rules used by ONCOdata for decision-making purpose were drawn from clinical guidelines. These guidelines do not make use of patient's biomedical information, so the decisions about treatments are made without taking individual issues not included in the clinical records into account. However, such additional information is considered by professionals as important for improving the quality and the safety of the care they deliver to the patients. This goal is addressed in this work by translational research methods.

The translational component in this work is the combination of ONCOdata with a biomedical system focused on the relation of genetic disorders and orthologous genes, namely, the OGO system. OGO does not only integrate information on genetic disorders but also provides orthology information that can be used for translational research. According to this project, we have integrated the bioinformatics repository, OGO, into the medical decision support system, ONCOdata, in order to provide such information that can justify the final decision made by doctors. The ONCOdata decision module can now provide better justification or even improve the knowledge of physicians on hereditary diseases and may suggest some extra medical tests on patients. Thus, ONCOdata decision is a more effective decision support tool now.

The genetic information is provided by the OGO knowledge base. That information was conceptualized and integrated using Semantic Web Technologies. That is to say, we have a formal repository based on description logics, which can be queried using semantic query languages, such as SPARQL. Hence, we extended the OGO system to provide query methods based on web technologies that allow non supervised applications to consult the OGO knowledge base. That knowledge base is built by retrieving and integrating information from a series of resources. The maintenance of such base is currently mainly manual, although methods for the formalization of the mappings between the resources and the global ontology used in the OGO system will facilitate a more automatic maintenance approach.

Web services are a technology solution based on the W3C standards, such as WSDL and SOAP, which provide a widespread and easy way to deploy services. So, it facilitates interoperability between systems and their technological independence. Such standards aim to describe the services and define the messages and the data formats that can be exchanged. The exploitation of the OGO data from ONCOdata was facilitated by the deployment of a series of web services. The web services permit to query the OGO KB by not only ONCOdata but also by other applications that implement their interfaces. Thus, Web services are focused on the loosely-coupled communication between applications.

Since query methods are encapsulated as services, applications are only connected at runtime and the relationships between them are minimal. In addition, interoperability between different software platforms is ensured by using web standard protocols such as SOAP, WSDL, and HTTP. Although, some of the web services were defined as a single string query, we have also developed a web service to query the OGO using their own defined SPARQL queries, which provides applications a more advanced query interface. Hence, applications can exploit the advantages of semantic repositories.

The implementation details of the OGO system have been encapsulated by the description of the services in order to ensure the availability and independence of them. In this way, we achieve that independent applications can query automatically the integrated information repository. For example, applications that support medical trials can take advantage of the type of information that OGO KB integrates when they seek for biological causes of genetic diseases.

Moreover, although the OGO system continues to develop and incorporating new features, the service descriptions can remain unchanged. Therefore, access to web services is easily achieved and requires little effort compared to the benefits. However, we plan to extend the ways in which third-party applications can exploit the OGO knowledge base, for instance, through REST services, which are widely used nowadays. Moreover, given that the OGO system is based on Semantic Web technologies, we plan to provide access to OGO via semantic web services.

6   Conclusions

In this paper, we have presented an improvement to the ONCOdata system through the incorporation of biomedical information for allocating medical treatments. Before, ONCOdata was not able to use genetic information related to the diseases managed by the system in order to suggest treatments for the patients. Now, such information is available, and can be also used by the physicians. Apart from the purely genetic information, the OGO system feeds ONCOdata with a series of links to scientific publications of interest for the physicians. Therefore, ONCOdata provides now a better support to physicians for selecting the best treatment for their patients. The results have been validated by the medical consultants of the company and this new option is included in the current version of ONCOdata.

In addition, encapsulating the complexity of Semantic Web Technologies for querying advanced systems such as OGO, facilitates a rapidly reutilization of systems. Therefore, applications can seek information from semantic repositories automatically without human supervision and also choose the level of complexity of queries they need. This has been achieved by providing services based on keywords and string based queries and services based on semantic languages such as SPARQL.

Acknowledgement. This work has been possible thanks to the Spanish Ministry for Science and Education through grant TSI2007-66575-C02-02 and the Spanish Ministry for Science and Innovation through grant TIN2010-21388-C02-02. Jose Antonio Miñarro has been supported by the Seneca Foundation and the Employment and Training Service through grant 07836/BPS/07.

References

1. Helmons, P.J., Grouls, R.J., Roos, A.N., Bindels, A.J., Wessels-Basten, S.J., Ackerman, E.W., Korsten, E.H.: Using a clinical decision support system to determine the quality of antimicrobial dosing in intensive care patients with renal insufficiency. Quality and Safety in Health Care 19(1), 22–26 (2010)
2. Kawamoto, K., Houlihan, C.A., Balas, E.A., Lobach, D.F.: Improving clinical practice using clinical decision support systems: a systematic review of trials to identify features critical to success. BMJ (2005); bmj.38398.500764.8F
3. Kaiser, K., Miksch, S.: Versioning computer-interpretable guidelines: Semi-automatic modeling of 'Living Guidelines' using an information extraction method. Artificial Intelligence in Medicine 46, 55–66 (2008)
4. Gadaras, I., Mikhailov, L.: An interpretable fuzzy rule-based classification methodology for medical diagnosis. Artificial Intelligence in Medicine 47, 25–41 (2009)
5. Kann, M.: Advances in translational bioinformatics: computational approaches for the hunting of disease genes. Brief Bioinform. 11(1), 96–110 (2010)
6. Knaup, P., Ammenwerth, E., Brandner, R., Brigl, B., Fischer, G., Garde, S., Lang, E., Pilgram, R., Ruderich, F., Singer, R., Wolff, A., Haux, R., Kulikowski, C.: Towards clinical bioinformatics: advancing genomic medicine with informatics methods and tools. Methods of Information in Medicine 43(3), 302–307 (2004)

7. Ruttenberg, A., Clark, T., Bug, W., Samwald, M., Bodenreider, O., Chen, H., Doherty, D., Forsberg, K., Gao, Y., Kashyap, V., Kinoshita, J., Luciano, J., Marshall, M., Ogbuji, C., Rees, J., Stephens, S., Wong, G.T., Wu, E., Zaccagnini, D., Hongsermeier, T., Neumann, E., Herman, I., Cheung, K.: Advancing translational research with the semantic web. BMC Bioinformatics 8, 2 (2007)
8. Berners-Lee, T., Hendler, J., Lassila, O.: The semantic web. Scientific American 284(5), 34–43 (2001)
9. Prokosch, H., Ganslandt, T.: Perspectives for medical informatics. reusing the electronic medical record for clinical research. Methods of Information in Medicine 48(1), 38–44 (2009)
10. Miranda-Mena, T., Benítez-Uzcategui, S., Ochoa, J., Martínez-Béjar, R., Fernández-Breis, J., Salinas, J.: A knowledge-based approach to assign breast cancer treatments in oncology units. Expert Systems with Applications 31, 451–457 (2006)
11. Goldblatt, E., Lee, W.: From bench to bedside: the growing use of translational research in cancer medicine. American Journal of Translational Research 2, 1–18 (2010)
12. Miñarro Gimenez, J., Madrid, M., Fernandez Breis, J.: Ogo: an ontological approach for integrating knowledge about orthology. BMC Bioinformatics 10(suppl. 10:S13) (2009)
13. Miranda-Mena, T., Martínez-Costa, C., Moner, D., Menárguez-Tortosa, M., Maldonado, J., Robles-Viejo, M., Fernśndez-Breis, J.: MIURAS 2: Motor de integración universal para aplicaciones sanitarias avanzadas. In: Inforsalud (2010)
14. Kang, B.: Validating Knowledge Acquisition: Multiple Classification Ripple Down Rules. PhD thesis, University of New South Wales (1996)
15. Miñarro-Gimenez, J., Madrid, M., Fernandez-Breis, J.: An integrated ontological knowledge base about orthologous genes and proteins. In: Proceedings of 1st Workshop SWAT4LS 2008, vol. 435 (2008)
16. Miñarro-Giménez, J.A., Aranguren, M.E., García-Sánchez, F., Fernández-Breis, J.T.: A semantic query interface for the OGO platform. In: Khuri, S., Lhotská, L., Pisanti, N. (eds.) ITBAM 2010. LNCS, vol. 6266, pp. 128–142. Springer, Heidelberg (2010)

MedFMI-SiR: A Powerful DBMS Solution for Large-Scale Medical Image Retrieval[*]

Daniel S. Kaster[1,2], Pedro H. Bugatti[2], Marcelo Ponciano-Silva[2],
Agma J.M. Traina[2], Paulo M.A. Marques[3], Antonio C. Santos[3],
and Caetano Traina Jr.[2]

[1] Department of Computer Science, University of Londrina, Londrina, PR, Brazil
dskaster@uel.br
[2] Department of Computer Science, University of São Paulo at São Carlos, SP, Brazil
{pbugatti,ponciano,agma,caetano}@icmc.usp.br
[3] Department of Internal Medicine – RPMS/University of São Paulo (USP) – Brazil
{pmarques,acsantos}@fmrp.usp.br

Abstract. Medical systems increasingly demand methods to deal with the large amount of images that are daily generated. Therefore, the development of fast and scalable applications to store and retrieve images in large repositories becomes an important concern. Moreover, it is necessary to handle textual and content-based queries over such data coupled with DICOM image metadata and their visual patterns. While DBMSs have been extensively used to manage applications' textual information, content-based processing tasks usually rely on specific solutions. Most of these solutions are targeted to relatively small and controlled datasets, being unfeasible to be employed in real medical environments that deal with voluminous databases. Moreover, since in existing systems the content-based retrieval is detached from the DBMS, queries integrating content- and metadata-based predicates are executed isolated, having their results joined in additional steps. It is easy to realize that this approach prevent from many optimizations that would be employed in an integrated retrieval engine. In this paper we describe the MedFMI-SiR system, which handles medical data joining textual information, such as DICOM tags, and intrinsic image features integrated in the retrieval process. The goal of our approach is to provide a subsystem that can be shared by many complex data applications, such as data analysis and mining tools, providing fast and reliable content-based access over large sets of images. We present experiments that show that MedFMI-SiR is a fast and scalable solution, being able to quickly answer integrated content- and metadata-based queries over a terabyte-sized database with more than 10 million medical images from a large clinical hospital.

1 Introduction

Health care institutions currently generate huge amounts of images in a variety of image specialties. Therefore, an efficient support is required to retrieve

---

[*] This work has been supported by CNPq, FAPESP, Capes and Microsoft Research.

C. Böhm et al. (Eds.): ITBAM 2011, LNCS 6865, pp. 16–30, 2011.

information from the large datasets accumulated. Picture Archiving and Communications Systems (PACS) are software that allow managing image distribution in a health care environment. A PACS encompasses an interface between the screening equipments and the workstations in which the exams are analyzed, handling images coded in the Digital Imaging and Communications in Medicine (DICOM) format. The DICOM format stores medical- and image collecting-related information together with each image. Although most routine tasks in medical applications search for images based on their associated metadata, several works have been showing that the retrieval based on the image visual characteristics can complement text-based search, opening new data analysis opportunities.

The concept of Content-Based Image Retrieval (CBIR) covers the techniques to retrieve images regarding their intrinsic information, such as visual patterns based on color, texture, shape and/or domain specific features. To find similar images, a CBIR system compares a given image with the images in the dataset according to a certain similarity criterion. It can be found in the literature several CBIR applications and techniques focused on medical data [21,1]. However, the majority of the approaches are not scalable to large datasets, being unfeasible to be employed in many real health care environments, which usually handle huge amounts of data.

Database Management Systems (DBMS) are usually employed to deal with large amounts of data. Nevertheless, existing DBMSs do not support the retrieval of complex data, such as images. As a consequence, when a CBIR system relies on a DBMS, the DBMS is used only to perform metadata-based queries, having the CBIR tasks executed in a separate engine. In such environment, answering queries combining metadata- and content-based operations requires joining the results of the two engines. Therefore, such queries cannot be optimized using query processing techniques neither in the CBIR side nor in the DBMS, as each engine is detached from the other. Thus, it is desirable to integrate the qualities of the CBIR systems and a DBMS into a unique application.

Other aspect that needs to be considered is that retrieval is just a part of a health care application. If this task is enclosed into a complex data management subsystem, it can employ highly specialized database strategies to speed up query processing. Moreover, it can take advantage of transaction control, backup and other fundamental operations that DBMSs already provide. This subsystem could serve different types of applications, from routine activities to knowledge discovery algorithms and decision support systems, in a controlled environment. Besides, to accomplish such integration there is no need of specific client libraries, since the CBIR core code is embedded into the DBMS.

This paper describes the MedFMI-SiR (Medical user-defined Features, Metrics and Indexes for Similarity Retrieval), which is a software module we developed associating the management of DICOM images by content with an Oracle Database, aimed at integrating the advantages of a DBMS, a PACS and a CBIR system. It allows storing huge amounts of textual DICOM metadata, visual information (intrinsic features) and medical images in an integrated way. Our approach allowed not only mixing the existing textual and visual information to achieve a better flexibility and accuracy regarding query answering, but also managing the communication of a great information flow, centralized in a unique application. We report experiences performed using more than 10 million images of several medical specialties from a large hospital, showing that our approach is very efficient, effective and scalable to perform both textual- and content-based image retrieval.

The remainder of this paper is structured as follows. Section 2 summarizes the main strategies used in medical data retrieval. Section 3 describes MedFMI-SiR, while Section 4 presents experiments performing the index creation and integrated queries, discussing the results achieved. Finally, Section 5 presents the conclusions and future directions.

2 Fundamental Concepts and Related Work

2.1 Content-Based Medical Image Retrieval

Medical information comprise a wide variety of data types, including text, time series and images. With the increasing amount of medical images being produced, it is necessary to provide ways to efficiently search them. In the last years, it can be noticed increasing interest in performing Content-Based Image Retrieval (CBIR) over medical data. The notion of CBIR is related to any technology that helps to manage digital images using their visual content [8]. An image is essentially a pixel matrix with values derived from sensors, whose content semantics is defined by the visual patterns pictured. Therefore, the first step to deal with an image by content is to generate a *feature vector*, or *signature*, that represent one or more of such visual patterns. The algorithms that produce the feature vectors are called *feature extractors*. There are several feature extractors for medical images in the literature. They are categorized generally as color-based (e.g. [9]), texture-based (e.g. [25]) or shape-based (e.g. [3]), and are usually focused on a specific kind of medical image.

Image feature vectors are complex data types, having the similarity among elements as the most employed concept to organize them. The similarity between two feature vectors is usually computed as the inverse of the distance between them in the embedded space. A distance is a function $\delta : \mathbb{S} \times \mathbb{S} \rightarrow \mathbb{R}^+$, where $\mathbb{S}$ is the domain of the feature vector. Therefore, the smaller the distance between two feature vectors the higher the similarity between the corresponding images. There are several distance functions in the literature. The most employed distances in CBIR systems are those from the L_p family (e.g. Euclidean, Manhattan or Chebychev). Nevertheless, the use of other measures to enhance similarity evaluation is quickly increasing, as is reported in [11].

Similarity queries are the fundamental retrieval operations over complex data. The main comparison operators employed to perform similarity queries are the *Range query* (Rq) and the *k-Nearest Neighbor query* (k-NNq) [7]. Both retrieve elements similar to a reference one ($s_q \in \mathbb{S}$), according to a similarity measure. An Rq aims at finding the elements dissimilar to s_q up to a certain maximum threshold given in the query. A k-NNq selects the k elements most similar to s_q. Such queries can be very helpful in a medical environment. For example, finding the cases most similar to the current one can help to improve the radiologist confidence or to train radiology residents. It can be found several approaches applying CBIR techniques to the medical field. Examples include systems for image retrieval and classification (e.g. [12,2]) and for enhancing PACS with CBIR techniques (e.g. [6,29]). Reviews of medical CBIR approaches can be found in [21,19,1].

2.2 Medical Image Retrieval Systems Combining Textual and Content-Based Information

Medical images commonly have descriptive text associated. Such metadata come from various sources, such as patient records and exam reports. When stored using the DICOM standard, the images themselves store additional information as metadata, as the DICOM format provides a comprehensive set of fields (tags) to describe the main information related to the image. The metadata are useful to help searching in medical image datasets, but there are limitations for their usability. For example, several important DICOM tags are manually filled by the acquisition machine operator, such as the admitting diagnoses and the study/series description. Therefore, they are subject to typographical errors, lacks of standardization, and there are tags whose content is subjective. Moreover, when the user does not know exactly what to look for, a situation that frequently occurs when dealing with medical data, the right query formulation becomes more challenging and error prone.

Combining text-based and content-based image retrieval can lead to a better accuracy, because one complements the other. There are several approaches to combine text- and content-based retrieval of medical images, including early and late fusion of weighted results from different search engines. For instance, the work described in [16] employs techniques to create automatic annotations that become part of the indexing process and are used as textual attributes to filter or reorder the results during the retrieval. Other works, like [22] and [20], allow users to provide both keywords and example images to define the desired result. Thereafter, these systems perform, in a transparent way, a textual and a content-based query and merge the partials to generate the final result returned to the user. However, in all these systems providing combined textual and content-based queries, the integration of the two types of searches is done in a separate step. This also occurs in other successful systems, such as the Image Retrieval in Medical Applications (IRMA) project [18] and the Spine Pathology & Image Retrieval System (SPIRS) [14]. Those systems do not take advantage of properties involving the filtering predicates, which in most situations would allow to enhance the overall performance, depending on the filter selectivities.

Analysing the existing medical image retrieval approaches, one can notice that most of them are targeted to relatively small and controlled datasets. Most of the works found in the literature were tested over image databases with up to a few thousand images. A few approaches dealt with larger databases storing little more than 100,000 images (e.g. [20], [14] and [26]). This can be explained by the fact that to evaluate the effectiveness of the proposed techniques it is necessary to employ controlled datasets, which are usually small. However, it is still lacking works over large image databases, evaluating the proposed techniques in an environment closer to real health care applications, where hundreds of images are generated every day, producing databases where millions of images are stored, each image with associated textual information. It is worth mentioning that the systems able to perform content-based medical image retrieval are essentially designed to be end-user applications. This conception limits the reuse of functionalities as building blocks for new applications, leading to an unnecessary effort to (re)codify similar functions in new systems.

DBMSs have a long history of success to manage huge amounts of data, but they need to be extended to provide native support for content-based medical image handling. Integrating such support into the DBMS core allows exploring optimizations that could be widely reused in new medical applications. The focus of this paper is to develop an operational support to meet the described needs and provide adequate solutions for those problems in a novel technique to combine metadata- and content-based conditions in a complex query to generate a unique query plan. This query plan can thus be manipulated using optimization techniques similar to those that have successfully been employed in the DBMS query processing engines.

2.3 DBMS-driven Medical Image Retrieval Approaches

Existing works that accomplish the integration of content- and metadata-based medical image retrieval usually rely on a database server only for searching over the metadata. With regard to DICOM data, software libraries are employed to extract the image headers, which are stored in DBMS tables. However, as the DICOM standard is complex and allows vendor-specific metadata tags, organizing the database is a laborious task. To alleviate this process, DBMS vendors have been developing functionalities to natively handle DICOM data, such as the Oracle Multimedia DICOM extension [24]. The DICOM support has been enhanced in the the recent Oracle DBMS versions, allowing increasing the security, integrity and performance of storing diagnostic images and other DICOM content.

Including content-based support in DBMS requires providing feature extractors and functions to evaluate the similarity between the generated feature vectors. Furthermore, the retrieval has to be fast, employing solutions that scale well with the size of the database. Several works have addressed algorithms and data structures for similarity retrieval [7,27]. For instance, the Slim-tree [30] is a dynamic Metric Access Method (MAM), which divides the space in regions determined by similarity relationships (distances) among the indexed elements. It enables pruning the subtrees that do not contain result candidates for a similarity query, which reduces disk accesses and distance calculations and, consequently, improves the search performance. Works can also be found on similarity query optimization, such as [31,5,28], addressing issues as how to include similarity operators in DBMS query processors.

Oracle Multimedia provides an extension to search images based on their content [23]. This extension enhances the database system with a new data type storing intrinsic image features and with indexes to speed up queries. In [10] is presented a web-based image retrieval system for medical images, which uses both the DICOM metadata and image content-based retrieval mechanisms provided by Oracle. The main drawbacks of the Oracle's image content-based mechanism are that there is only one feature extraction method, which is not specialized for the medical field, and it is closed-source, disallowing enhancements. Similarly, early versions of the IBM DB2 database server provided native content-based retrieval [15], but this support was also closed-source and without support for medical images.

There also are open source systems to perform content-based search over images. One example is the SIREN (SImilarity Retrieval ENgine), which is a prototype that implements an interpreter over a DBMS such as Oracle and PostgreSQL [4]. It recognizes an extension to the SQL language that allows the representation of similarity queries. However, the SIREN does not handle DICOM images and, as it is a blade over the database system, queries are executed using two complementary plans, one for executing the content-based search and the other for executing the remainder SQL operators. Another example is the PostgreSQL-IE (PostgreSQL with Image-handling Extension) [13]. The PostgreSQL-IE is an extension to the PostgreSQL DBMS that encapsulates the images in a new data type and provides a number of feature extractors for medical images. However, it does not support images in the DICOM format. Moreover, as it is implemented through user-defined functions, the content-based retrieval functions are not treated as first-order database operators and therefore they are not touched during the query optimization process.

The next section presents the MedFMI-SiR, which is a database module for medical image retrieval that allows evaluating alternative query plans involving metadata-based as well as content-based predicates and is targeted to very large databases.

3   The MedFMI-SiR System

This section describes the MedFMI-SiR (Medical user-defined Features, Metrics and Indexes for Similarity Retrieval). It is a powerful, extensible and scalable DBMS solution that allows combining metadata- and content-based retrieval of medical images.

3.1   MedFMI-SiR Architecture

The MedFMI-SiR is a module attached to the Oracle Database that allows managing DICOM images by content integrated to the DBMS features. It is an extension of the FMI-SiR module [17] to handle medical images. Fig. 1 illustrates the MedFMI-SiR architecture.

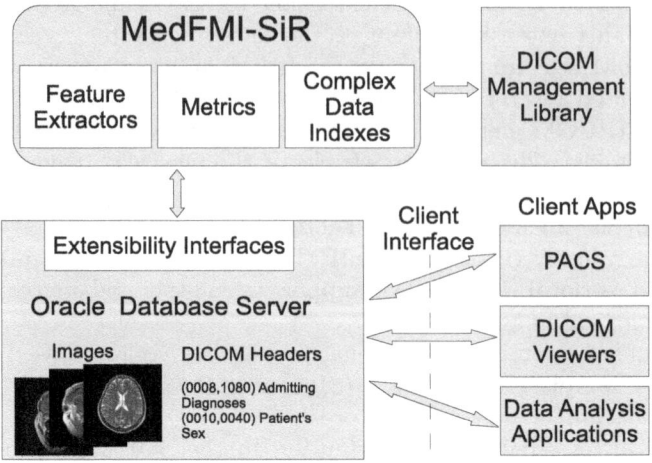

Fig. 1. The MedFMI-SiR architecture

The MedFMI-SiR is attached to the DBMS using the Oracle's extensibility interfaces, providing the image retrieval functionalities to the client applications in a transparent way. It interacts with the DICOM Management Library, which is the DICOM Toolkit (DCMTK)[1] in the current implementation. The DCMTK is a collection of C++ libraries and applications implementing the DICOM standard, including software to examine, construct and convert image files into/from the DICOM format. The MedFMI-SiR is capable of opening several modalities of DICOM images, including compressed images, to perform image processing operations. The feature extractors provided generate feature vectors from the DICOM content, which are stored in regular database columns. The MedFMI-SiR also allows performing queries combining metadata- and content-based conditions, providing metrics and complex data indexes. It is controlled by the DBMS query processor, thus providing a tight integration to other DBMS operations. The index capabilities are achieved integrating the Arboretum[2] library through the Oracle extensibility indexing interface, providing an efficient processing of similarity queries.

Client applications interact with the module accessing directly the DBMS using SQL, without requiring any additional software libraries. Therefore, the MedFMI-SiR approach can serve multiple applications, such as PACS, DICOM viewers and other already existing data analysis applications. Moreover, as it is open source, it can be enriched with domain specific feature extractors and metrics, regarding different image modalities, improving the retrieval quality. Feature extractors and metrics, as well as indexes, are organized into portable C++ libraries, which makes easier to develop new functionalities.

---

[1] http://dicom.offis.de/dcmtk.php.en
[2] http://gbdi.icmc.usp.br/arboretum

3.2   DICOM Metadata Management

The MedFMI-SiR metadata handling is twofold: it uses the DICOM Management Library, if any metadata is necessary during feature extraction; and it uses the native Oracle DICOM support, for storage and retrieval purposes. This section focuses on the retrieval mechanism, which is provided by the Oracle Database. The Multimedia DICOM extension was introduced in the Oracle Database to allow it to manage the enterprise medical data repository, serving as the storage and retrieval engine of PACS and physician workstations, as well as other hospital information systems. Within the Oracle's extension, DICOM data are handled using a special data type, which reads the header metadata and stores them in XML format together with the raw image.

For illustration purposes, in this paper we consider that the DICOM images are stored in a table created as follows:

```
CREATE TABLE exam_images (
  img_id INT PRIMARY KEY,
  dicom_img ORDDICOM NOT NULL);
```

where `ORDDICOM` is the Oracle's proprietary data type to store DICOM images. When DICOM content is stored in the database, it can be manipulated like other relational data using SQL. Therefore, medical applications are able to take advantage of the features provided by the DBMS back end, such as concurrent operation, access control, integrity enforcing and so forth. The Oracle's XML native support also allows mixing XML and regular statements in SQL queries. For instance, the following query selects the identifiers of the studies whose images are labeled with "Multiple Sclerosis" admitting diagnosis (the admitting diagnoses are stored as DICOM tag 0008,1080)[3].

```
SELECT DISTINCT e.dicom_img.getStudyInstanceUID()
FROM exam_images e
WHERE extractValue(e.dicom_img.metadata,
  tag["00081080"]) = 'Multiple Sclerosis';
```

The function `extractValue` scans the XML data and identifies the elements that satisfy the given expression. Metadata-based queries can be efficiently processed using a function-based index, which creates a B-tree index on the `extractValue` function result. The following statement creates a B-tree index on the function result, employed to speedup the previous query.

```
CREATE INDEX diagnoses_ix ON exam_images e (
  extractValue(e.dicom_img.metadata,  tag["00081080"]);
```

3.3   Image Feature Vector Generation

As MedFMI-SiR is attached to the database server, it is possible to employ the loading methods of the DBMS, such as bulk insertions and direct-path load,

---

[3] The queries in this paper employ a simplified syntax to improve readability.

which are much more efficient than individual insertions. This feature is very desirable in real environments, as it has been noticed that existing CBIR systems usually employ much more costly operations to build the database. After having loaded the binary data, it is necessary to extract both the DICOM metadata, which is performed using the Oracle functions, and the image features. The DICOM image feature extraction is performed calling a new proxy function that we developed, called generateSignature, in a SQL command. The SQL compiler maps it to a C++ function, which calls the target algorithm in the feature extractor library. The extractor scans the DICOM image and returns the corresponding feature vector. The feature vector is serialized and stored in the database as a Binary Large OBject (BLOB) attribute. For example, the following command can be issued to define a relation to store the feature vectors:

```
CREATE TABLE exam_signatures (
  img_id INT PRIMARY KEY REFERENCES exam_images(img_id),
  img_histogram BLOB);
```

where img_histogram is an attribute that stores feature vectors. Regarding this schema, the features of all images stored in the exam_images relation can be extracted and stored issuing the following SQL statements:

```
FOR csr IN (
  SELECT i.dicom_img, s.img_histogram
  FROM exam_images i, exam_signatures s
  WHERE i.img_id = s.img_id FOR UPDATE) LOOP
    generateSignature('', csr.dicom_img, csr.img_histogram, 'Histogram');
END LOOP;
```

where Histogram is the desired feature extractor, the attribute dicom_img from relation exam_images stores the images to have the features extracted, and the attribute img_histogram is an output parameter, employed to store the generated feature vector. As the generateSignature function modifies the database state, the system must access the data through an exclusive lock (in the example indicated by the FOR UPDATE clause).

The MedFMI-SiR also supports including additional feature extractors as well as loading features extracted by external software, if they can be stored in a text file. This functionality is enabled filling the first parameter of the generateSignature function with the file name holding the features. In this case, the last two parameters are ignored and the BLOB signature is populated with the features read from the file.

3.4   Metadata- and Content-Based Search Integration

After having the feature vectors computed, the images are ready to be compared by similarity. Similarity queries are based on distance functions, which are defined in MedFMI-SiR as follows:

```
<distance_name>_distance(signature1 BLOB, signature2 BLOB);
```

The distance function returns the distance between the two features (signatures) as a real value. Several distance functions are available, such as `Manhattan_distance`, `Euclidean_distance` and `Canberra_distance`. Moreover, other distance functions can also be included easily. The evaluation of a similarity query in MedFMI-SiR starts when the application poses a SQL query to the database manager and the query processor identifies one of the similarity operators that we developed. A range query is written as in the following example:

```
SELECT img_id FROM exam_signatures
WHERE Manhattan_dist(img_histogram,
  center_histogram) <= 0.5;
```

where `center_histogram` is a BLOB containing the query center feature vector, the relational operator `<=` indicates that a range query has been requested, and the value `0.5` is the range threshold. In a similar way, a k-NN query with $k = 10$ is stated as follows:

```
SELECT img_id FROM exam_signatures
WHERE Manhattan_knn(img_histogram,
  center_histogram) <= 10;
```

The MedFMI-SiR provides new index types based on dynamic MAMs to improve the performance of similarity queries, for instance, the Slim-tree (see Section 2.3). The index creation for images follows the same syntax used for regular indexes.

The tight integration of MedFMI-SiR and Oracle allows posing queries mixing traditional and metadata-related conditions and content-based queries. Moreover, queries are written using the SQL standard and they are submitted to the DBMS using regular programming language drivers. For instance, the following query identifies the 50 images whose histograms are the most similar to the given reference, and from them returning the images referring to series described as "Routine Epilepsy Protocol" (the series description is stored in the DICOM tag 0008,103E).

```
SELECT e.dicom_img
FROM exam_images i, exam_signatures s
WHERE i.img_id = s.img_id
 AND Manhattan_knn(s.img_histogram,
   center_histogram) <= 50
 AND extractValue(i.dicom_img.metadata,
   tag["0008103E"]) = 'Routine Epilepsy Protocol';
```

Therefore, the MedFMI-SiR approach makes possible to exploit alternative query plans and to efficiently execute queries through indexed searching over both image similarities and metadata predicates. The query processing enhancement can be particularly perceived in large databases and in data-intensive tasks, such as data mining algorithms and operations with complex predicates involving many similarity and metadata conditions over several tables. The next section presents results of experiments over a very large database of images from real cases from a university hospital.

4 Experimental Evaluation

We evaluated MedFMI-SiR over several medical image datasets. In this section we present the results obtained using a real dataset composed of 14 million of DICOM images acquired in the Clinical Hospital of Ribeirão Preto – USP, stored in a table occupying 1.7TB of disk space. Those images correspond to about 18 months of the past hospital activities. The tests were executed on a Oracle Database 11g Enterprise Edition Release 2 under a Ubuntu Server GNU/Linux 9.10 64bits, running on a machine equipped with an Intel Core 2Quad 2.66GHz processor, 4GB of RAM and four 7,200rpm SATA2 1TB disks.

For this experiment, we extracted the 256-bin gray-level histogram from each image. The average extraction rate was 102 histograms per second, producing a table with 13.4GB for the features. We used the gray-level histogram in these tests because, initially, we are interested only in showing that our proposed techniques are able to scale up to the full amount of data and not yet in targeting any specific medical specialty. In fact, although its identification property is small, gray-level histograms are one of the most common techniques used to represent images for every specialty. Although extracting the histograms of all those past images took approximately 28 hours of processing, processing the new images from the daily hospital routine is now be much faster and unnoticed by the health staff, as each image has its histogram extracted as soon as it is stored in the database.

4.1 Index Creation

In this section we present the experiments performed to evaluate the scalability of the indexes, that is, the index behavior as the database increases. To this purpose, we randomly sampled the original features to create tables with sizes of 1,000, 10,000, 100,000, 1,000,000 and 10,000,000 tuples. For each table, we created a Slim-tree index on the feature attribute and performed a round of fifty 50-NN queries with randomly selected query references, aimed at evaluating the index search performance. Fig. 2 shows the results obtained by such experiments measuring: (a) the time to create the indexes, (b) the disk space required for the index, and (c) the average time required to perform each 50-NN query, respectively. Analyzing them, we can clearly notice that the growth rate presented a linear behavior according to the table size. Since both axes in the graphs are in log scale and the slopes of the plots are up to one, the corresponding algorithms have linear cost. These results confirm that the MedFMI-SiR indexes are scalable.

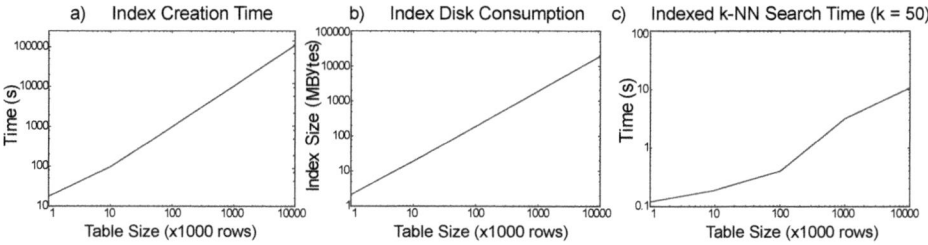

Fig. 2. Behavior of the MedFMI-SiR regarding the scalability of: (a) index creation time, (b) requirement of disk space for indexing, and (c) time to answer 50-NN queries

4.2   Query Processing

The MedFMI-SiR approach allows posing both metadata-based, content-based and integrated queries over medical images. One of the main advantages of the MedFMI-SiR's architecture is to allow generating alternative plans, enabling the DBMS to perform on-demand query rewriting. Therefore, in this section we present the results obtained when combining k-NN queries with metadata from the DICOM images. To this intent, we employed the last query presented in Section 3.4 and executed k-NN queries for k varying from 1 to 200. For each value of k, we took the average of 50 queries with random centers.

We forced this query to be executed in MedFMI-SiR following two alternative query plans, shown in figures 3(a) and 3(b). The query plan shown on Fig. 3(a) executes an indexed scan over the signature table using the k-NN similarity predicate and joins its results to the corresponding rows that satisfy the metadata predicate in the image table. The query plan shown on Fig. 3(b) executes an indexed scan over the signature table and another indexed scan over the image table using the metadata tag, joining the results. In these plans, the query processor employed the available indexes on the tables' primary keys to speed up the join operation. To complete the evaluation, we also generated another query plan, which first executes an indexed scan over the image table regarding the metadata predicate and then computes the k-NN over the resulting joined rows, as illustrated in Fig. 3(c). Notice that this plan answers a query that is different from the one answer by plans (a) and (b). In this case, the query returns the k images closest to the query center from those whose series description is "Routine Epilepsy Protocol".

Fig. 4 shows the execution times regarding these query plans, varying the value of k and the selectivity of the metadata predicate. The metadata condition of the query corresponding to Fig. 4(a) has a selectivity of about 99.95%, being satisfied by a little more than 5,000 tuples, and the metadata condition of the query corresponding to Fig. 4(b) has a selectivity of approximately 95%, being satisfied by around 500,000 tuples. It can be seen in Fig. 4(a) that the two first query plans (`knn-ind-meta-seq` and `knn-ind-meta-ind`) present sub-linear behaviors as k grows, although the second plan is up to 32% faster. The query plan `knn-seq-meta-ind` is faster than the others, which can be explained by the

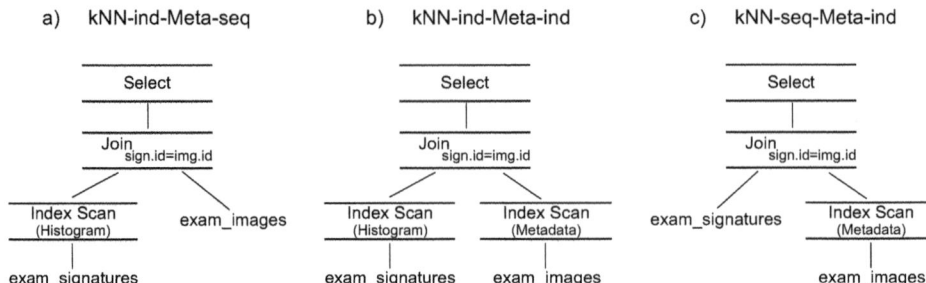

Fig. 3. (a, b) Alternative query plans for the last query of Section 3.4. (c) A query plan for a slightly different query, executing an indexed scan on the metadata predicate followed by an in-memory k-NNq.

high selectivity of the metadata predicate. Moreover, it executes in almost constant time, since the filtering predicate does not depend on k. In Fig. 4(b) the behaviors of the query plans are the opposite, as the selectivity of the metadata condition is lower. It can be seen in this figure that the query plans that employ an indexed search on the metadata condition (i.e. `knn-ind-meta-ind` and `knn-seq-meta-ind`) have a very poor performance, requiring several minutes to execute the query. Moreover, the execution time of the plan `knn-ind-meta-ind` have a very slight variation regarding the value of k, because the cost of the k-NN indexed search is dominated by the cost of the indexed metadata search. On the other hand, the plan `knn-ind-meta-seq` answers the query in few seconds, being up to two orders of magnitude faster than the other plans (notice that the graph of Fig. 4(b) is in log scale). This experiment illustrates the importance of allowing enumerating alternative query plans, according to the selectivities of the predicates and the query operators employed, which is a feature provided by the MedFMI-SiR.

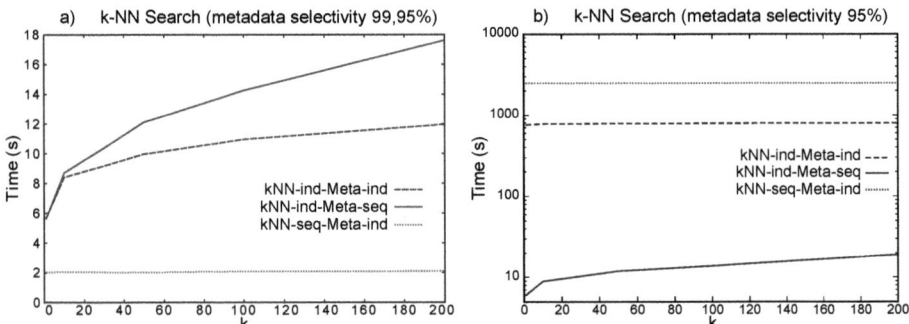

Fig. 4. Comparing execution times in FMI-SiR using different query plans. (a) Query combining a 50-NN search and a metadata filter with selectivity 99.95%. (b) Query combining a 50-NN search and a metadata filter with selectivity 95%.

5   Conclusion

In this paper we presented the MedFMI-SiR (Medical user-defined Features, Metrics and Indexes for Similarity Retrieval), an open source module that executes similarity queries over DICOM images integrating metadata- and content-based image retrieval. Our solution is attached onto the Oracle DBMS, benefiting from all the capabilities it natively provides. There is no need of specific client libraries, since the all code is embedded into the DBMS, allowing a centralized installation. Our module also allows adding new feature extraction algorithms, distance functions as well as index structures. The results obtained show that the presented approach is fast and scalable, and provides highly flexibility to pose queries integrating metadata- and content-based retrieval of medical images over very large datasets. Future works include employing other feature extractors to explore features tailored to selected medical specialties.

References

1. Akgül, C., Rubin, D., Napel, S., Beaulieu, C., Greenspan, H., Acar, B.: Content-based image retrieval in radiology: Current status and future directions. J. Digital Imaging 24, 208–222 (2011)
2. Alto, H., Rangayyan, R.M., Desautels, J.E.L.: Content-based retrieval and analysis of mammographic masses. Journal of Electronic Imaging 14(2), 1–17 (2005)
3. Balan, A.G.R., Traina, A.J.M., Ribeiro, M.X., Marques, P.M.D.A., Traina Jr., C.: HEAD: the human encephalon automatic delimiter. In: CBMS 2007, Maribor, Slovenia, pp. 171–176. IEEE, Los Alamitos (2007)
4. Barioni, M.C.N., Razente, H.L., Traina, A.J.M., Traina, C.J.: SIREN: A similarity retrieval engine for complex data. In: VLDB 2006, Seoul, South Korea, pp. 1155–1158. ACM, New York (2006)
5. Berchtold, S., Böhm, C., Keim, D.A., Krebs, F., Kriegel, H.P.: On optimizing nearest neighbor queries in high-dimensional data spaces. In: Van den Bussche, J., Vianu, V. (eds.) ICDT 2001. LNCS, vol. 1973, pp. 435–449. Springer, Heidelberg (2000)
6. Bueno, J.M., Chino, F.J.T., Traina, A.J.M., Traina Jr., C., Marques, P.M.d.A.: How to add content-based image retrieval capability in a PACS. In: CBMS 2002, Maribor, Slovenia, pp. 321–326. IEEE, Los Alamitos (2002)
7. Böhm, C., Berchtold, S., Keim, D.A.: Searching in high-dimensional spaces - index structures for improving the performance of multimedia databases. ACM Computing Surveys 33(3), 322–373 (2001)
8. Datta, R., Joshi, D., Li, J., Wang, J.Z.: Image retrieval: Ideas, influences, and trends of the new age. ACM Computing Surveys 40(2), 1–60 (2008)
9. Deselaers, T., Keysers, D., Ney, H.: Features for image retrieval: An experimental comparison. Information Retrieval 11(2), 77–107 (2008)
10. Dimitrovski, I., Guguljanov, P., Loskovska, S.: Implementation of web-based medical image retrieval system in oracle. In: ICAST, pp. 192–197. IEEE, Los Alamitos (2009)
11. Felipe, J.C., Traina Jr., C., Traina, A.J.M.: A new family of distance functions for perceptual similarity retrieval of medical images. J. Digital Imaging 22(2), 183–201 (2009)
12. Greenspan, H., Pinhas, A.T.: Medical image categorization and retrieval for PACS using the GMM-KL framework. IEEE Trans. on Inf. Technology in Biomedicine 11(2), 190–202 (2005)
13. Guliato, D., Melo, E.V., Rangayyan, R.M., Soares, R.C.: PostgreSQL-IE: An image-handling extension for PostgreSQL. J. Digital Imaging 22(2), 149–165 (2008)

14. Hsu, W., Antani, S., Long, L.R., Neve, L., Thoma, G.R.: SPIRS: A web-based image retrieval system for large biomedical databases. International Journal of Medical Informatics 78(1), 13–24 (2009)
15. IBM Corp.: Image, audio, and video extenders administration and programming guide, DB2 Universal Database Version 8 (2003)
16. Kalpathy-Cramer, J., Hersh, W.: Multimodal medical image retrieval: image categorization to improve search precision. In: MIR 2010, Philadelphia, Pennsylvania, USA, pp. 165–174. ACM, New York (2010)
17. Kaster, D.S., Bugatti, P.H., Traina, A.J.M., Traina Jr., C.: FMI-SiR: A flexible and efficient module for similarity searching on Oracle database. JIDM 1(2), 229–244 (2010)
18. Lehmann, T.M., Güld, M., Thies, C., Fischer, B., Spitzer, K., Keysers, D., Ney, H., Kohnen, M., Schubert, H., Wein, B.B.: Content-based image retrieval in medical applications. Methods of Informatics in Medicine 43, 354–361 (2004)
19. Long, L.R., Antani, S., Deserno, T.M., Thoma, G.R.: Content-based image retrieval in medicine: Retrospective assessment, state of the art, and future directions. IJHISI 4(1), 1–16 (2009)
20. Müller, H., Deselaers, T., Deserno, T.M., Kalpathy–Cramer, J., Kim, E., Hersh, W.: Overview of the imageCLEFmed 2007 medical retrieval and medical annotation tasks. In: Peters, C., Jijkoun, V., Mandl, T., Müller, H., Oard, D.W., Peñas, A., Petras, V., Santos, D. (eds.) CLEF 2007. LNCS, vol. 5152, pp. 472–491. Springer, Heidelberg (2008)
21. Müller, H., Michoux, N., Bandon, D., Geissbuhler, A.: A review of content-based image retrieval systems in medical applications-clinical benefits and future directions. Int. Journal of Medical Informatics 73(1), 1–23 (2004)
22. Névéol, A., Deserno, T.M., Darmoni, S.J., Güld, M.O., Aronson, A.R.: Natural language processing versus content-based image analysis for medical document retrieval. J. of the American Society for Information Science and Technology 60(1), 123–134 (2009)
23. Oracle Corp.: Oracle interMedia User's Guide, 10g Release 2 (10.2) (2005)
24. Oracle Corp.: Oracle Multimedia DICOM Developer's Guide, 11g Release 2 (2009)
25. Pereira-Jr., R.R., de Azevedo-Marques, P.M., Honda, M.O., Kinoshita, S.K., Engelmann, R., Muramatsu, C., Doi, K.: Usefulness of texture analysis for computerized classification of breast lesions on mammograms. Journal of Digital Imaging 20(3), 248–255 (2007)
26. Rahman, M.M., Antani, S.K., Thoma, G.R.: A classification-driven similarity matching framework for retrieval of biomedical images. In: MIR 2010, Philadelphia, Pennsylvania, USA, pp. 147–154. ACM, New York (2010)
27. Samet, H.: Foundations of Multidimensional and Metric Data Structures. Morgan Kaufmann, San Francisco (2006)
28. Silva, Y.N., Aref, W.G., Ali, M.H.: The similarity join database operator. In: ICDE 2010, Long Beach, California, USA, pp. 892–903. IEEE, Los Alamitos (2010)
29. Tan, Y., Zhang, J., Hua, Y., Zhang, G., Huang, H.: Content-based image retrieval in picture archiving and communication systems. In: Medical Imaging 2006: PACS and Imaging Informatics, San Diego, CA, USA, vol. 6145, pp. 282–289. SPIE, San Jose (2006)
30. Traina Jr., C., Traina, A.J.M., Faloutsos, C., Seeger, B.: Fast indexing and visualization of metric datasets using Slim-trees. IEEE Trans. on Knowl. and Data Eng. 14(2), 244–260 (2002)
31. Traina Jr., C., Traina, A.J.M., Vieira, M.R., Arantes, A.S., Faloutsos, C.: Efficient processing of complex similarity queries in RDBMS through query rewriting. In: CIKM 2006, Arlington, VA, USA, pp. 4–13. ACM, New York (2006)

Novel Nature Inspired Techniques in Medical Information Retrieval

Miroslav Bursa[1], Lenka Lhotská[1], Vaclav Chudacek[1], Michal Huptych[1], Jiri Spilka[1], Petr Janku[2], and Martin Huser[2]

[1] Department of Cybernetics,
Faculty of Electrical Engineering,
Czech Technical University in Prague, Czech Republic
[2] Obstetrics and Gynaecology Clinic,
University Hospital in Brno, Czech Republic

Abstract. In this work we have studied, evaluated and proposed different swarm intelligence techniques for mining information from loosely structured medical textual records with no apriori knowledge. We describe the process of mining a large dataset of ∼50,000–120,000 records × 20 attributes in DB tables, originating from the hospital information system recording over 10 years. This paper concerns only textual attributes with free text input, that means 613,000 text fields in 16 attributes. Each attribute item contains ∼800–1,500 characters (diagnoses, medications, etc.). The output of this task is a set of ordered/nominal attributes suitable for rule discovery mining.

Information mining from textual data becomes a very challenging task when the structure of the text record is very loose without any rules. The task becomes even harder when natural language is used and no apriori knowledge is available. The medical environment itself is also very specific: the natural language used in textual description varies with the personality creating the record, however it is restricted by terminology (i.e. medical terms, medical standards, etc.). Moreover, the typical patient record is filled with typographical errors, duplicates and many (nonstandard) abbreviations.

Nature inspired methods have their origin in real nature processes and play an important role in the domain of artificial intelligence. They offer fast and robust solutions to many problems, although they belong to the branch of approximative methods. The high number of individuals and the decentralized approach to task coordination in the studied species revealed high degree of parallelism, self-organization and fault tolerance.

First, classical approaches such as basic statistic approaches, word (and word sequence) frequency analysis, etc., have been used to simplify the textual data and provide an overview of the data. Finally, an ant-inspired self-organizing approach has been used to automatically provide a simplified dominant structure, presenting structure of the records in the human readable form that can be further utilized in the mining process as it describes the vast majority of the records.

Note that this project is an ongoing process (and research) and new data are irregularly received from the medical facility, justifying the need for robust and fool-proof algorithms.

C. Böhm et al. (Eds.): ITBAM 2011, LNCS 6865, pp. 31–38, 2011.
© Springer-Verlag Berlin Heidelberg 2011

Keywords: Swarm Intelligence, Ant Colony, Text Mining, Data Mining, Medical Record Processing, Hospital Information System.

1   Introduction

In many industrial, business, healthcare and scientific areas we witness the boom of computers, computational appliances, personalized electronics, high-speed networks, increasing storage capacity and data warehouses. Therefore a huge amount of various data is transferred and stored, often mixed from different sources, containing different data types, unusual coding schemes, and seldom come without any errors (or noise) and omissions.

Even with rapidly increasing computational power of modern computers, the analysis of huge databases becomes very expensive, making the development of novel techniques reasonable. Especially in text processing, the impact of automated methods is crucial. In contrary to classical methods, nature-inspired methods offer many techniques, that can increase speed and robustness of classical methods.

Nature inspired metaheuristics play an important role in the domain of artificial intelligence, offering fast and robust solutions in many fields (graph algorithms, feature selection, optimization, clustering, etc). Stochastic nature inspired metaheuristics have interesting properties that make them suitable to be used in data mining, data clustering and other application areas.

Plenty of nature inspired methods are studied and developed in present. One category is represented by methods, that are inspired by the behavior of ant colonies. These methods have been applied to many problems (often NP-hard). Review can be seen in [4] and [1]. We concentrate on the state-of-the-art nature methods inspired by the social behavior of insect communities, by the swarm intelligence, brain processes and other real nature processes.

Ant colonies inspired many researches to develop a new branch of stochastic algorithms: *ant colony inspired algorithms*. Based on the ant metaphor, algorithms for both static and dynamic combinatorial optimization, continuous optimization and clustering have been proposed. They show many properties similar to the natural ant colonies, however, their advantage lies in incorporating the mechanisms, that allowed the whole colonies to effectively survive during the evolutionary process.

Cemetery formation and brood sorting are two prominent examples of insects' collective behavior. However, other types of ant behavior have been observed, for example predator-prey interaction, prey hunting, etc. The most important are mentioned below.

The accuracy for relation extraction in journal text is typically about 60 % [5]. A perfect accuracy in text mining is nearly impossible due to errors and duplications in the source text. Even when linguists are hired to label text for an automated extractor, the inter-linguist disparity is about 30 %. The best results are obtained via an automated processing supervised by a human [7].

2   Input Dataset Overview

The dataset consists of a set of approx. 50 to 120 thousand records (structured in different relational DB tables) × approx. 20 attributes. Each record in an attribute contains about 800   to 1500 characters of text (diagnoses, patient state, anamneses, medications, notes, references to medical stuff, etc.). For textual mining, 16 attributes are suitable, providing us with approx 613,000 text fields.

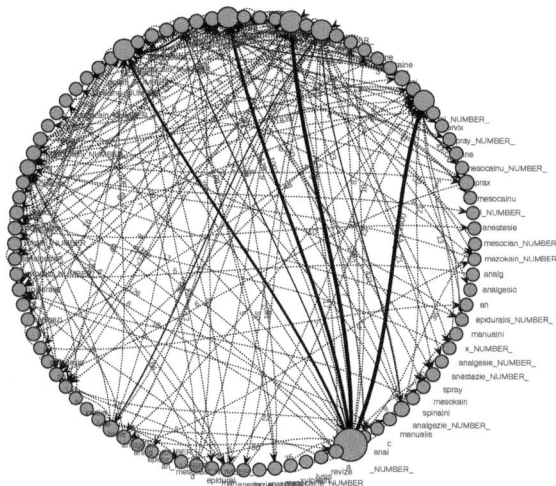

Fig. 1. Figure shows a transitional diagram (directed graph) structure of single attribute literals (a subset). Circular visualization has been used to present the amount of literal transitions (vertices).

The overview of one small (in field length) attribute is visualized in Fig. 1. Only a subsample (about 5 %) of the dataset could be displayed in this paper, as the whole set would render into an incomprehensible black stain. The vertices (literals) are represented as a green circle, the size reflects the literal frequency. By literal we mean a separate word. Edges represent transition states between literals (i.e. the word flow in a record (sentence)); edge stroke shows the transition rate (probability) of the edge. The same holds for all figures showing the transition graph (i. e. Fig. [5, 3, 4, 6]) only different visualization approach has been used. The transition graphs show the most frequent sentence flows contained in the record and are further used for rules specification that retrieve specific information from the record of an attribute.

Fig. 2 depicts the potence of the attributes: the count of vertices and edges (note that the y-axis has different scales). It is clear, that human interpretation and analysis of the textual data is very fatiguing, therefore any computer aid is highly welcome.

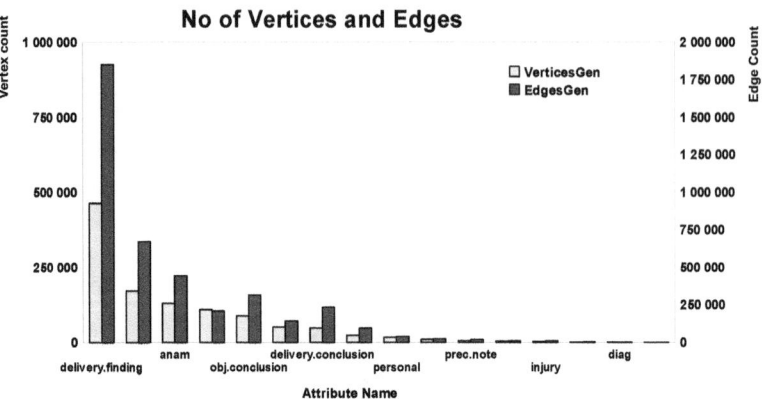

Fig. 2. The DB contains 16 textual attributes that are susceptible for information retrieval via natural language literal extraction. Number of literals (vertices) and transitions (edges) in the probabilistic models are shown for each attribute.

3   Nature Inspired Techniques

Social insects, i. e. ant colonies, show many interesting behavioral aspects, such as self-organization, chain formation, brood sorting, dynamic and combinatorial optimization, etc. The coordination of an ant colony is of local nature, composed mainly of indirect communication through pheromone (also known as *stigmergy*, the term has been introduced by Grassé et al. [6]), although direct interaction communication from ant to ant (in the form of antennation) and direct communication have also been observed [9].

The Ant Colony Optimization (ACO) [4] is a metaheuristic approach, inspired by the ability of ants to discover shortest path between nest and food source. The process is guided by deposition of a chemical substance (pheromone). As the ants move, they deposit the pheromone on the ground (amount of the pheromone deposited is proportional to the quality of the food source discovered). Such pheromone is sensed by other ants and the amount of pheromone changes the decision behavior of the ant individual. The ant will more likely follow a path with more pheromone.

This approach has been utilized (with improvements) to simplify the structure of the vast dataset (Fig. 1) by finding the most important state transitions between literals, producing a probabilistic transitional model. The output structure (Fig. 3) is presented to the analyst for further processing/iteration. The ant-colony clustering is used to create groups of similar literals. Only the Hamming metrics has been used so far for measuring the similarity, however the Levenshtein distance or cosine metrics is much appropriate.

For clustering, the ACO_DTree method [3,2] and ACO inspired clustering [8] variations have been successfully used.

Fig. 3. An automatically Ant-Colony induced transition graph showing the most important relations in one textual attribute. The ACO approach has been used to cluster the corresponding vertices. Refer to section 2.

The basic idea of ACO can be described as follows:

```
a1    Repeat until stopping criterion is reached
a2       Create ants
a3       Construct solutions
a4       Evaporate Pheromone
a5       Daemon Actions (pheromone deposit)
a6    End Repeat
```

New solutions are constructed continuously. It is a stochastic decision process based on the pheromone amount sensed by the ants. As in nature, the pheromone slowly evaporates over time (over iterations) in order to avoid getting stuck in local minimum and to adapt to dynamically changing environment. Daemon actions represent *background* actions which consist mainly of pheromone deposition. The amount is proportional to the quality of solution (and appropriate adaptive steps).

Main parameters of the algorithm are (with major importance to the method proposed): pheromone lay rate, pheromone evaporate rate, number of solutions created (number of ants), number of iterations, etc.

Ant Inspired Clustering: [8] method belongs to the group of nature inspired methods (together with neural networks, self-organizing maps, swarm intelligence, evolutionary computing, etc.). It is a stochastic metaheuristic, which uses a similarity measure and the paradigm of ant species, and which is able to collect similar items (broods, dead bodies) together.

The model can be described as follows: First the data vectors are randomly scattered onto a two-dimensional grid (usually a toroidal one). Ants (agents) are then randomly placed onto the grid. In each iteration step an ant searches its neighborhood and computes a probability of picking up a vector (if the ant is unloaded and steps onto a vector) or a probability of dropping down a vector (if the ant is loaded and steps onto a free grid element). The probability of picking up a vector is higher when an ant steps on a vector that introduces heterogeneity in its surrounding. The probability of dropping down the vector is higher when an ant carrying a vector steps into an area of similar vectors. These rules lead to formation of clusters in the 2D grid which can be easily obtained.

4 Automated Processing

Automated layout of transition graph is very comfortable, however the contents of the attribute is so complicated, that a human intervention is inevitable. Examples of automated layout can be seen in Fig. 3 and Fig. 4. In Fig. 3 only ant-colony clustering method has been used to lay out vertices and edges. An improvement based on text position (literal distance from record start) has been implemented. The result can be seen in Fig. 4. It is clear that the second figure is much more comprehensible and self-explanatory.

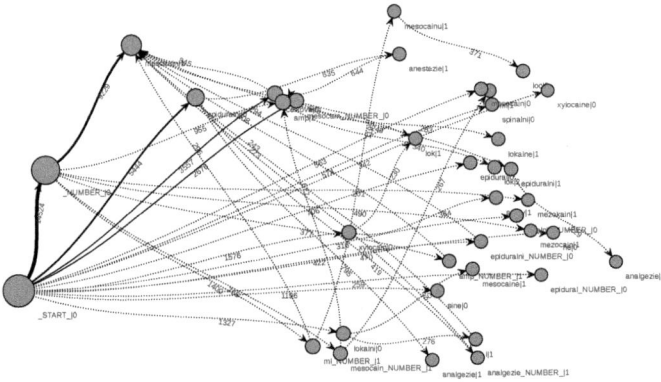

Fig. 4. A fully automated transition graph showing the most important relations in one textual attribute. The ACO approach has been used to cluster the corresponding vertices. Refer to section 2.

5 Expert Intervention

A human intervention and supervision over the whole project is indiscutable. Therefore also human (expert) visualization of the transition graph has been studied.

An example of expert layout is presented in Fig. 5. The vertices are (usually) organized depending on the position in the text (distance from the starting point) as these have the highest potence. Number literal (a wildcard) had the highest potence, as many quantitative measures are contained in the data (age, medication amount, etc.). Therefore it has been fixed to the following literal, spreading into the graph via multiple nodes. This allowed to organize the chart visualization in more logical manner. Time needed to organize such graph was about 5–10 minutes.

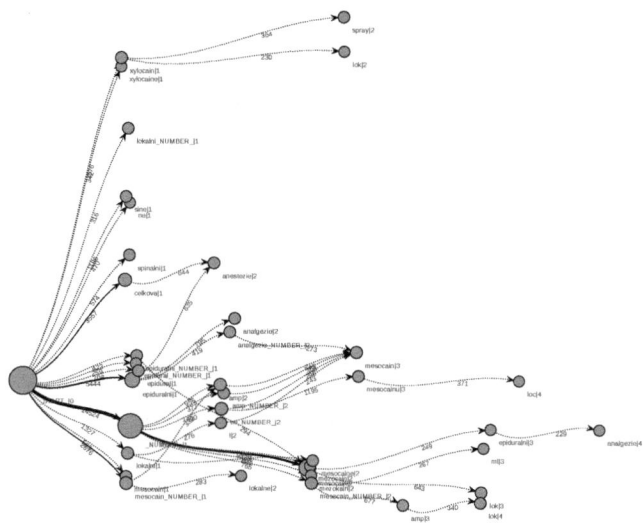

Fig. 5. An expert (human) organized transition graph showing the most important relations in one textual attribute. Refer to section 2

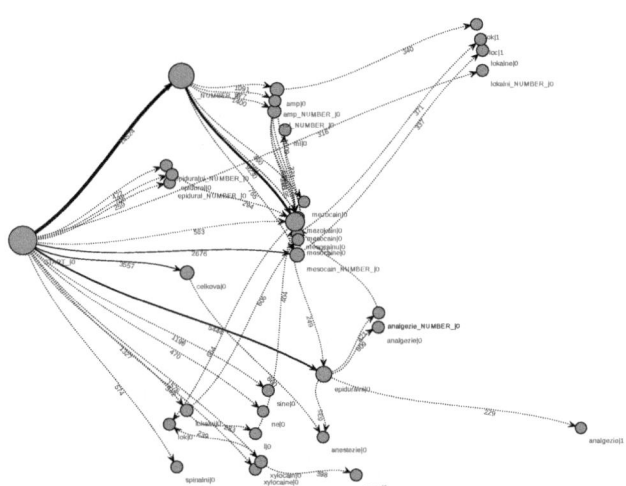

Fig. 6. A semi-automated (corrected by a human expert) organized transition graph showing the most important relations in one textual attribute. Refer to section 2

An aid of a human expert has been used in semi-automated approach (see Fig. 6) where the automated layout has been corrected by the expert. The correction time has been about 20–30 seconds only.

6   Results and Conclusion

The main advantage of the nature inspired concepts lies in automatic finding relevant literals and groups of literals that can be adopted by the human analysts and furthermore improved and stated more precisely. The use of induced probabilistic models in such methods increased the speed of loosely structured textual attribute analysis and allowed the human analysts to develop lexical analysis grammar more efficiently in comparison to classical methods. The speedup (from about 5–10 minutes to approx 20–30 seconds) allowed to perform more iterations, increasing the yield of information from data that would be further processed in rule discovery process. However, the expert intervention in minor correction is still inevitable.

Acknowledgements. This project has been supported by the research programs MSM 6840770012 "Transdisciplinary Research in the Area of Biomedical Engineering II" of the CTU in Prague, sponsored by the Ministry of Education, Youth and Sports of the Czech Republic and by the project NT11124-6/2010 "Cardiotocography evaluation by means of artifficial intelligence" of the Ministry of Health Care. This work has been developed in the BioDat research group http://bio.felk.cvut.cz.

References

1. Blum, C.: Ant colony optimization: Introduction and recent trends. Physics of Life Reviews 2(4), 353–373 (2005)
2. Bursa, M., Huptych, M., Lhotska, L.: Ant colony inspired metaheuristics in biological signal processing: Hybrid ant colony and evolutionary approach. In: Biosignals 2008-II, vol. 2, pp. 90–95. INSTICC Press, Setubal (2008)
3. Bursa, M., Lhotska, L., Macas, M.: Hybridized swarm metaheuristics for evolutionary random forest generation. In: Proceedings of the 7th International Conference on Hybrid Intelligent Systems, pp. 150–155. IEEE CSP, Los Alamitos (2007)
4. Dorigo, M., Stutzle, T.: Ant Colony Optimization. MIT Press, Cambridge (2004)
5. Freitag, D., McCallum, A.K.: Information extraction with hmms and shrinkage. In: Proceedings of the AAAI Workshop on Machine Learining for Information Extraction (1999)
6. Grasse, P.-P.: La reconstruction du nid et les coordinations inter-individuelles chez bellicositermes natalensis et cubitermes sp. la theorie de la stigmergie: Essai d'interpretation des termites constructeurs. Insectes Sociaux 6, 41–81 (1959)
7. Lafferty, J., McCallum, A., Pereira, F.: Conditional random fields: Probabilistic models for segmenting and labeling sequence data. In: Proceedings of the ICML, pp. 282–289 (2001); Text processing: interobserver agreement among linquists at 70
8. Lumer, E.D., Faieta, B.: Diversity and adaptation in populations of clustering ants. In: From Animals to Animats: Proceedings of the 3rd International Conference on the Simulation of Adaptive Behaviour, vol. 3, pp. 501–508 (1994)
9. Trianni, V., Labella, T.H., Dorigo, M.: Evolution of direct communication for a swarm-bot performing hole avoidance. In: Dorigo, M., Birattari, M., Blum, C., Gambardella, L.M., Mondada, F., Stützle, T. (eds.) ANTS 2004. LNCS, vol. 3172, pp. 130–141. Springer, Heidelberg (2004)

Combining Markov Models and Association Analysis for Disease Prediction

Francesco Folino and Clara Pizzuti

Institute for High Performance Computing and Networking (ICAR)
National Research Council of Italy (CNR)
Via Pietro Bucci, 41C
87036 Rende (CS), Italy
{ffolino,pizzuti}@icar.cnr.it

Abstract. An approach for disease prediction that combines clustering, Markov models and association analysis techniques is proposed. Patient medical records are clustered and a Markov model for each cluster is generated to perform prediction of illnesses a patient could likely be affected in the future. However, when the probability of the most likely state in the Markov models is not sufficiently high, it resorts to sequential association analysis, by considering the items induced by high confidence rules generated by recurring sequential disease patterns. Experimental results show that the combination of different models enhances predictive accuracy and is a feasible way to diagnose diseases.

1 Introduction

In recent years the definition of new health care models to determine the risk for individuals to develop specific diseases [17] is attracting a lot of interest among researchers. The prediction of illnesses based on the past patient medical history revealed efficacious in foreseeing diseases a patient could likely be affected in the future [5, 4, 8, 7, 18]. Progress in such a field is very important since it could significantly improve efficiency and effectiveness of health care strategies and have practical implications in the life quality of individuals.

Markov models [15] are well known and commonly used prediction models which have been employed with success for predicting the browsing behavior of users on the Web [6, 11–13]. There is a strong similarity between Web page access prediction and patient disease prediction. In fact the sequence of web pages accessed by a user can be considered as the sequence of diseases diagnosed to a patient along his life. Thus, advances obtained in the former context can be exploited in the medical setting in order to improve disease prediction accuracy.

In [7] a system, named *CORE*, that combines clustering and association analysis on the data set of patient records to generate local specialized and accurate prediction models has been presented. *CORE* uses the past patient medical history for generating models able to determine the risk of individuals to develop future diseases. The models are built by using the set of frequent diseases that contemporarily appear in the same patient. A patient is represented by the set of

C. Böhm et al. (Eds.): ITBAM 2011, LNCS 6865, pp. 39–52, 2011.

ICD-9-CM codes of diagnosed diseases, and a disease is predicted by comparing a patient with individuals having a similar clinical history.

In this paper we propose an extension of the approach presented in [7], named $CORE^+$, by introducing Markov models in order to improve prediction accuracy. $CORE^+$, analogously to $CORE$, clusters patient medical records, but then builds a Markov model for each cluster and performs prediction by using these models. However, when the probability of the most probable state in the Markov models is not sufficiently high (this concept will be formally explained in section 3), it resorts to sequential association analysis [19] by considering the items induced by high confidence rules generated by recurring sequential disease patterns. The medical record of a patient is thus compared with the patterns discovered by the models, and a set of illnesses is predicted.

$CORE^+$ differs from $CORE$ in two main aspects. The first is the introduction of Markov models, the second one is that it uses sequential pattern mining, which takes into account the order of diseases in a patient medical record. The approach has been inspired by Khalil et al. [13] and adapted for the medical context. However there is a main difference in the way association rules are applied. $CORE^+$ computes association rules for the cluster the patient belongs to, while Khalil et al. find association rules only for the Markov states for which a prediction has not a sufficiently high probability.

Experimental results show that the integration of Markov models increases the prediction accuracy of $CORE^+$ with respect to $CORE$, and it is a promising method to predict individual diseases by taking into account only the illnesses a patient had in the past.

The paper is organized as follows. The next section describes the concepts of clustering, Markov models, and association rules. In section 3 the prediction method is described. Section 4 reports the evaluation of the proposed approach on a data set of patient medical records. Section 5 concludes the paper.

2   Background Concepts

Before giving the details of the proposed approach, in this section the related concepts necessary to describe it are reported. Let m be the number of patients contained in the data set of patient histories. From this data set a new data set $T = \{t_1, \ldots, t_m\}$, where each t_i is a patient medical record of variable size constituted by a sequence of ICD-9-CM disease codes, is generated. Thus T represents the medical histories relative to all the m patients.

2.1   Clustering

Grouping the set T of patients in k groups having similar disease history is an effective way of improving the accuracy of the prediction model, as showed in [7]. Performing clustering is not, however, an easy task at all, since its performances are tightly related to the kind of method used. For our purposes, a variant of the traditional k-means clustering [10], able to deal with categorical tuples of variable size, is used.

For a given parameter k, this algorithm partitions T into k clusters $C = \{C_1, \ldots C_k\}$ such that high intra-cluster similarity and low inter-cluster similarity are guaranteed. Each record $t_i \in T$ is assigned to a cluster C_j according to its distance $d(t_i, r_j)$ from a vector r_j that represents the cluster at hand, called the *representative* or *center* of the cluster, with the aim to minimize a cost function. The cost function is computed as the sum of squared distances of each record from its cluster center. The algorithm works as follows. First of all, k records are selected from T randomly. They represent the initial cluster centers, and each other $t_i \in T$ is assigned to the closer cluster. Then, the algorithm updates the representative of each cluster and re-assigns each record consequently. The iterations terminate when the representatives do not change any more.

It is worth noting that the above schema is parametric w.r.t. the definitions of distance d and representative r. Since in our scenario we deal with categorical data, we used a kind of distance that proved to work very well in this setting: the *Jaccard* distance. This measure is derived by the *Jaccard coefficient* [1, 9] which is based on the idea that the similarity between two itemsets is directly proportional to the number of their common items and inversely proportional to the number of different ones. Therefore, given two records t_i and $t_j \in T$, the Jaccard distance can be defined as:

$$d(t_i, t_j) = 1 - \frac{|t_i \cap t_j|}{|t_i \cup t_j|} \tag{1}$$

The next step pertains a suitable definition for the cluster representatives. An effective way for building the representative consists in using the frequent items belonging to the cluster. The frequency degree can be controlled by introducing a user-defined threshold value γ representing the minimum percentage of occurrences an item must have for being inserted into the cluster representative. More formally, given $T_{C_i} = \{t_1, \ldots, t_q\}$ the set of records belonging to the cluster C_i, $D_{C_i} = \bigcup_i t_i = \{d_1, \ldots, d_p\}$ the set of items of C_i, and $\gamma \in [0, 1]$, then the representative r_{C_i} for C_i can be computed as follows:

$$r_{C_i} = \{d \in D_{C_i} | f(d, T_{C_i})/q \geq \gamma\} \tag{2}$$

where $f(d, T_{C_i}) = |\{t_i | d \in T_{C_i}\}|$.

The clustering algorithm assumes that the number of clusters k has to be fixed at the beginning. Thus, another open issue is how to set k in order to obtain the best partitioning. Ideally, the best partitioning is achieved for the value k^* in correspondence of which the cost function has its global minimum. However, find k^* could be unfeasible in practice. Therefore, we pragmatically recurred to a sub-optimal solution: we iterated the clustering algorithm by ranging k in $[1, |T|]$ until the first, local minimum is reached.

2.2   Association Analysis

Employing association analysis [2, 3] for prediction purposes is not new in the data mining literature. It relies on the concept of *frequent itemsets* to extract strong correlations among the items constituting the data set to study. Originally, association analysis has been applied to market basket data, where each item represents the purchase done by a customer. However, it can be easily transposed into the medical context by associating an item with a disease, and by considering an itemset as the set of diseases a patient had along his life until the present. For extracting sequential patterns, we apply the well-known *PrefixSpan* algorithm [16] that efficiently searches for sequential patterns by cutting the exponential search space of candidate itemsets. The concept of frequency is formalized through the concept of *support*.

Typically, groups of diseases occurring frequently together in many transactions are referred to as *frequent itemsets*. Given a set $I_{C_i} = \{i_1, \ldots, i_l\}$ of frequent itemsets induced on a cluster $C_i = \{t_1, \ldots, t_p\}$, the support of an itemset $i_j \in I$, $\sigma(i_j)$, is defined as:

$$\sigma(i_j) = \frac{\mid \{t_i \mid i_j \subseteq t_i, t_i \in C_i\} \mid}{\mid C_i \mid} \tag{3}$$

where $\mid . \mid$ denotes the number of elements in a certain set. The support, thus, determines how often a group of diseases appear together. It is a very important measure because very low support discriminates those groups of items occurring only by chance. Thus a frequent itemset, to be considered interesting, must have a support greater than a fixed threshold value *minsup*.

An association rule is an implication expression of the form $X \Rightarrow Y$, where X and Y are disjoint itemsets. The importance of an association rule is measured by both its *support* and *confidence* values. The support of a rule is computed as the support of $X \cup Y$ and tells how often a rule is applicable. The confidence is defined as $\sigma(X \cup Y)/\sigma(X)$, and determines how frequently items in Y appear in transactions that contain X. Frequent itemsets having a support value above a minimum threshold are used to extract high confidence rules that can be exploited to build a prediction model, by matching the medical record of a patient against the patterns discovered by the model.

In our scenario, the support determines how often a group of diseases appears together, while a rule like $X \Rightarrow \{d\}$ (where X is a set of frequent diseases and d is a new disease) having a high confidence, allows to reliably infer that d will appear along with the diseases contained in X.

2.3   Markov Models

Markov models are a well known technique for understanding stochastic processes and have been extensively used as prediction models because of the good accuracy levels they reach. Deshpande and Karypis [6] represent Markov models as a triple $< A, S, TPM >$, where A is a set of actions, S is a set of states, and $TPM = \mid S \mid \times \mid A \mid$ is a transition probability matrix, where an entry $tpm_{i,j}$ is the probability that the action j is performed when the process is in the state i. The simplest Markov model, known as *first-order Markov model*, predicts the next action by looking only to the previous action. In general, a *w-order Markov model* makes predictions by considering the last w actions. In the context of predicting user's web behavior, Deshpande and Karypis [6] identify the input data for building the Markov models as web sessions, i.e. the sequence of pages accessed by a user during a visit to a site. Thus, the actions are the pages of the web site, and the states are the w consecutive web pages observed in different sessions. In our medical context, the actions are the ICD-9-CM disease codes, the web sessions are the set $T = \{t_1, \ldots, t_m\}$ of patient medical records, and the states are the w $\{t_{i1}, \ldots, t_{iw}\}$ consecutive disease codes observed in T. The transition matrix TPM is then computed by counting how many times the code in position j appears after the state i.

t_1	$\{401, 715, 722, 723\}$
t_2	$\{401, 437, 715, 722, 756\}$
t_3	$\{401, 437, 715, 722, 723\}$
t_4	$\{437, 715, 722, 756\}$

Fig. 1. A set of four patient medical records

1st Order	401	437	715	722	723	756
$s_1 = 401$	0	2	1	0	0	0
$s_2 = 437$	0	0	3	0	0	0
$s_3 = 715$	0	0	0	4	0	0
$s_4 = 722$	0	0	0	0	2	2
$s_5 = 723$	0	0	0	0	0	0
$s_6 = 756$	0	0	0	2	0	0

Fig. 2. 1st order Transition Probability Matrix corresponding to the patient medical records of Figure 1

For example, consider the set of patient medical records reported in Figure 1. The first-order Transition Probability Matrix (Figure 2) is such that each state is constituted by a disease code, thus there are 6 different states. The entry in position $(2,3)$ is 3 because there are three medical records (t_2, t_3, t_4) in which the code 715 appears after the state $s_2 = 437$. Thus, the probability that the disease 715 be predicted as next disease after 437 is 1. The second-order TPM (Figure 3) contains the couples of codes appearing in sequence. Thus for the state $\{715, 722\}$ and the disease code 723, the entry of the matrix in position $(4,5)$ is 2 because 723 appears twice $(t_1$ and $t_3)$ after the couple $\{715, 722\}$. After the TPM of a fixed order is computed, given a sequence of ICD-9-CM codes, it is sufficient to lookup the TPM and extract the disease having the highest frequency. Given, for example, $\{756, 715\}$, by using the 1st-order TPM, the code predicted is 722.

A main drawback of Markov models is that, in order to obtain good predictive accuracy, higher-order models must be used, but higher-order models are computing demanding because of the high number of states that can be generated.

2nd Order	401	437	715	722	723	756
{401, 715}	0	0	0	1	0	0
{401, 437}	0	0	2	0	0	0
{437, 715}	0	0	0	3	0	0
{715, 722}	0	0	0	0	2	2
{722, 723}	0	0	0	0	0	0
{722, 756}	0	0	0	0	0	0

Fig. 3. 2nd order Transition Probability Matrix corresponding to the patient medical records of Figure 1

In the next section we show that the combination of low order Markov models and association rules improves prediction accuracy, keeping moderate runtime requirements.

3   A Hybrid Framework for Disease Prediction

Given a data set $T = \{t_1, \ldots, t_m\}$ of medical records, and fixed the order w of Markov models, the number k of clusters, the threshold γ for computing the cluster representatives, the support σ, and the confidence τ to compute association rules, the construction of the prediction models performs the following steps:

- *Cluster medical records in k clusters $\{C_1, \ldots, C_k\}$*
- *Build a Markov model $MM_{C_i}^w$ of order w for each cluster C_i*
- *Compute association rules for each cluster with support σ and confidence τ*

Once the models are built for each discovered cluster, if a new medical record is given, the prediction phase encompasses three main steps:

- *Cluster Assignment*, where the patient is recognized as member of a cluster C_i by matching him against each cluster representative;
- *Model Selection*, where the model M_i (relative to the corresponding cluster) is selected;
- *Prediction*, where the proper prediction is performed by exploiting M_i.

The *Prediction* step works in this way. We use a window of size w over the medical records for capturing the patient history depth used for the prediction. The size w means that we apply a Markov model of order w, thus only the last (in time order) w diseases appearing in the record influence the computation of possible forthcoming illnesses. If the Markov model prediction produces either a no state or a state having a not enough high probability (see below), the association rules are used instead. In this case, we consider the frequent itemsets of size $w + 1$ induced on C_i that contain the w items appearing in the current medical patient record $t_i \in C_i$. The prediction of the next disease is based on the confidence of the corresponding association rule whose antecedent are the w

frequent items of t_i, and the consequent is exactly the disease to be predicted. If this rule has a confidence value greater than a fixed threshold, its consequent is added to the set of predicted illnesses.

For the sake of clarity, let us perform a prediction on a medical patient record $t_i^w = \{t_1^i, \ldots, t_w^i\} \in C_i$ of size w. We first apply the wth Markov model learned on the cluster C_i. If t_i^w matches a state in this model, the next disease d_i is estimated by

$$\Pr(t_{w+1}^i) = \underset{d_i \in D_{C_i}}{\operatorname{argmax}}\{\Pr(t_{w+1}^i = d_i | t_w^i, t_{w-1}^i, \ldots, t_1^i)\}$$

d_i is accepted as the most probable next disease only if it results in a state whose probability is significantly better than that of the second most probable predicted disease.

More in detail, the $100(1 - \alpha)$ percent confidence interval around the most probable next disease is computed, and thus it is checked if the second predicted disease falls within this interval [6]. If this condition happens, the most probable state is discarded, otherwise it is accepted as next predicted disease. If $\hat{p} = \Pr(d_i)$ is the probability of the most probable disease, then its $100(1 - \alpha)$ percent confidence interval is given by

$$\hat{p} - z_{\alpha/2} * \sqrt{\frac{\hat{p}(1 - \hat{p})}{n}} \leq p \leq \hat{p} + z_{\alpha/2} * \sqrt{\frac{\hat{p}(1 - \hat{p})}{n}} \tag{4}$$

where $z_{\alpha/2}$ is the upper $\alpha/2$ percentage point of the standard normal distribution and n is the frequency of the Markov state [14].

In the case the Markov model is unable to provide us with a reliable prediction, the approach proposed in [7] is used instead to circumvent the problem. In particular, the patient medical record t_i^w is matched against all the frequent itemsets $I_{C_i}^{w+1}$ of size $w + 1$ induced on C_i. Each itemset $i_i^{w+1} \in I_{C_i}^{w+1}$ containing t_i^w contributes to the set of the candidate diseases with a prediction d_i. It is easy to note that $i_i^{w+1} = t_i^w \cup \{d_i\}$. Finally, if the confidence of the rule $t_i^w \Rightarrow \{d_i\}$ (i.e., $\sigma(t_i^w \cup \{d_i\})/\sigma(t_i^w)$) is greater than a fixed threshold τ, the disease d_i is considered reliable, and it is added to the set of predicted diseases.

Example. In order to explain the way our prediction approach works in practice, let us consider the set T of patient medical records reported in Table 1.

Let us also suppose $k = 2$ be the number of clusters the clustering algorithm is forced to find, and $\gamma = 0.5$ be the minimum percentage of occurrences a disease must have for being inserted into the cluster representative (see Equation 2). On the base of the above parameters, it is easily verifiable that the clustering algorithm finds the clusters C_1 and C_2, as reported in Tables 2 and 3, respectively. Furthermore, the clusters are equipped with their representatives: $r_{C_1} = \{401, 437, 592, 715, 722, 723, 756\}$ and $r_{C_2} = \{241, 255, 272, 595, 780\}$. After the clusters have been built, the hybrid disease prediction model is carried out for each cluster found.

Table 1. Set T of patient records involving some common diseases

t_1	$\{401, 715, 722, 723\}$
t_2	$\{401, 437, 715, 722, 756\}$
t_3	$\{437, 592, 715, 722, 723\}$
t_4	$\{401, 437, 592, 715, 722, 723\}$
t_5	$\{437, 715, 722, 756\}$
t_6	$\{592, 715, 722, 723, 756\}$
t_7	$\{401, 592, 715, 722, 756\}$
t_8	$\{401, 437, 715, 721, 722, 723\}$
t_9	$\{241, 255, 595, 780\}$
t_{10}	$\{241, 255, 272, 595, 780\}$

Table 2. Cluster C_1

t_1	$\{401, 715, 722, 723\}$
t_2	$\{401, 437, 715, 722, 756\}$
t_3	$\{437, 592, 715, 722, 723\}$
t_4	$\{401, 437, 592, 715, 722, 723\}$
t_5	$\{437, 715, 722, 756\}$
t_6	$\{592, 715, 722, 723, 756\}$
t_7	$\{401, 592, 715, 722, 756\}$
t_8	$\{401, 437, 715, 721, 722, 723\}$

Table 3. Cluster C_2

t_8	$\{241, 255, 595, 780\}$
t_9	$\{241, 255, 272, 595, 780\}$

Now, let $t = \{t_1 = 715, t_2 = 722\}$ be a new patient disease record. Since the distance $d(t, r_{C_i}) = 1 - 2/7 = 0.714$ is lower than $d(t, r_{C_2}) = 1$, t is recognized belonging to C_1, thus the Markov model built upon C_1 is exploited to perform the predictions. In particular, if the window size w is set to 2, the 2nd order Markov model is used. It is easy to note that the state $\{715, 722\}$ appears 7 times, and the diseases 723 and 756 appear 4 times and 3 times after this state, respectively, thus:

$$Pr(t_3) = \text{argmax}\{Pr(t_3 = 723 | t_2 = 722, t_1 = 715)\} = \text{argmax}\{t_3 = 723 \rightarrow 0.57\}$$

and

$$Pr(t_3) = \text{argmax}\{Pr(t_3 = 756 | t_2 = 722, t_1 = 715)\} = \text{argmax}\{t_3 = 756 \rightarrow 0.43\}$$

However, this information does not necessarily provide us with the the correct prediction of the next disease since there is not an enough probability difference for the diseases 723 and 756. More in detail, if we compute the confidence interval for the most probable next disease at 90% confidence level (i.e., $z_{\alpha/2} = 1.65$), we obtain that this may vary approximately between 0.33 and 0.85. Notice that, because of the small number of instances in this example, the greater the confidence level, the larger the confidence interval. Since the probability of the other disease $\Pr(t_3 = 756) = 0.43$ falls in this interval, we cannot consider the prediction made by the Markov model reliable. In this case of uncertainty, in order to disambiguate the choice, we resort to the predictive capability of the association rules in the same way we proposed in [7, 8]. By fixing $\sigma = 0.5$, the frequent itemsets for C_1 are shown in Table 4.

Table 4. Frequent itemsets built upon cluster C_1

I^1	I^2	I^3
{401} (5)	{401, 715} (5)	{401, 715, 722} (5)
{437} (5)	{401, 722} (5)	{437, 715, 722} (5)
{592} (4)	{592, 715} (4)	{592, 715, 722} (4)
{715} (8)	{437, 715} (5)	{715, 722, 723} (5)
{722} (8)	{437, 722} (5)	{715, 722, 756} (4)
{723} (5)	{715, 722} (8)	
{756} (4)	{715, 723} (5)	
	{715, 756} (4)	
	{722, 723} (5)	
	{722, 756} (4)	
	{592, 722} (4)	

By matching the medical record $t = \{715, 722\}$ against the 3-frequent itemsets I^3, both the diseases having codes 723 and 756 are candidate for being the likely, next diseases the patient t may incur in. However, the diseases 723 and 756 changes their status from candidate to predicted only if the confidence of the association rule $r1$: $\{715, 722\} \Rightarrow \{723\}$ and $r2$: $\{715, 722\} \Rightarrow \{756\}$ are greater than the minimum confidence threshold τ. If we set $\tau = 0.6$, since the confidence of $conf(r1) = \sigma(\{715, 722, 723\})/\sigma(\{715, 722\}) = 5/8 = 0.625$, and the $conf(r2) = \sigma(\{715, 722, 756\})/\sigma(\{715, 722\}) = 4/8 = 0.5$, the disease 723 is definitively added to the set of predicted illnesses. Therefore, by means of the rule $r1$, we foresee that a patient presenting osteoarthrosis (715), and disc disorders (722), he is very likely to present the symptom of other disorders of cervical region (723).

In the next section we show that $CORE^+$ is effective in predicting diseases.

4 Experimental Results

In this section we first define the measures used to test the effectiveness of our approach. Next, we present the results and evaluate them on the base of the introduced metrics. The dataset T we used for the experiments consists of 1105 patient records involving 330 distinct diseases. Each record contains, for each patient, the list of disease codes with the date of the visit in which that disease has been diagnosed. The disease codes are those defined by the International Classification of Diseases, Ninth Revision, Clinical Modification (ICD-9-CM).

In order to perform a fair evaluation, we applied the well-known 10-*fold cross validation* method [19], i.e., the original dataset is split in 10 equal-sized partitions. During each of the 10 runs, one of the partitions is chosen for testing, while the rest of them are used for training the prediction model. The cumulative error is found by summing up the errors for all the 10 runs. The strategy we followed for testing our approach is detailed in the following.

First of all, the records in the training set T_{train} are partitioned in k clusters, and for each group, a distinct prediction model M_i is built upon. Relatively to the dataset at hand, we empirically found that $k = 10$ and $\gamma = 0.5$ is the setting that ensures the best partitioning possible for the dataset at hand. A record t in the test set T_{test} is first assigned to one of the k cluster, then it is divided in two subsets of diseases. The first subset, called $head_t$, is used for generating predictions, while the remaining one, referred as $tail_t$, is used to evaluate the prediction. Actually, the length of $head_t$ is tightly related to the maximum window size w allowable for each cluster, and, intuitively, must be lower than the maximal length of frequent itemsets mined in each cluster. For instance, in this very specific case, since we verified that the prediction models M built on clusters (also for low values of support σ) produce frequent patterns of size at most 5, the maximum length of $head_t$ can't exceed 4. More in general, given a window size w, we select the first w diseases as $head_t$ and the remaining $|t| - w$ as $tail_t$. If the record t belongs to the cluster C_i, the relative prediction model M_i first applies the wth Markov model $MM_{C_i}^w$. If either $head_t$ does not match any state of $MM_{C_i}^w$ or its prediction has a low probability, the association rules for predicting the next diseases are exploited. To this purpose, $head_t$ is matched against all sequential patterns $I_{C_i}^{w+1}$. Note that, as explained in the previous section, low probability means that the second predicted disease falls within the confidence interval of the first predicted disease [6]. To compute the confidence interval we used a value of $z_{\alpha/2} = 1.65$. This value has been obtained by properly tuning it on the data set.

Let P_{head_t} be the set containing all the candidate predictions made by either exploting Markov Models or association rules. In this latter case, fixed the minimum confidence threshold τ, P_{head_t} will contain all the candidate predictions whose confidence is greater than τ. Subsequently, the set P_{head_t} is compared with $tail_t$. The comparison of these sets is done by using two different metrics, namely *Precision* and *Recall* [19]. Precision and recall are two widely used statistical measures in the data mining field. In particular, precision is seen as a measure of exactness, whereas recall is a measure of completeness.

By customizing these definitions to our scenario, we exploited precision for assessing how accurate the provided predictions are (i.e., the proportion of relevant predictions to the total number of predictions) and recall for testing if we predicted all the diseases the patients are likely to be affected in the future (i.e, the proportion of relevant predictions to all diseases that should be predicted). Formally, the precision of $P(head_t, \tau)$ is defined as:

$$Precision(P_{head_t}) = \frac{|P_{head_t} \cap tail_t|}{|P_{head_t}|}$$

and the recall of P_{head_t} as:

$$Recall(P_{head_t}) = \frac{|P_{head_t} \cap tail_t|}{|tail_t|}$$

The cumulative precision (recall) scores drawn in Figure 4 are computed as the mean of the precision (recall) values achieved by each single record $t \in T_{test}$ over the size of T_{test}. More in detail, we measured both precision and recall by

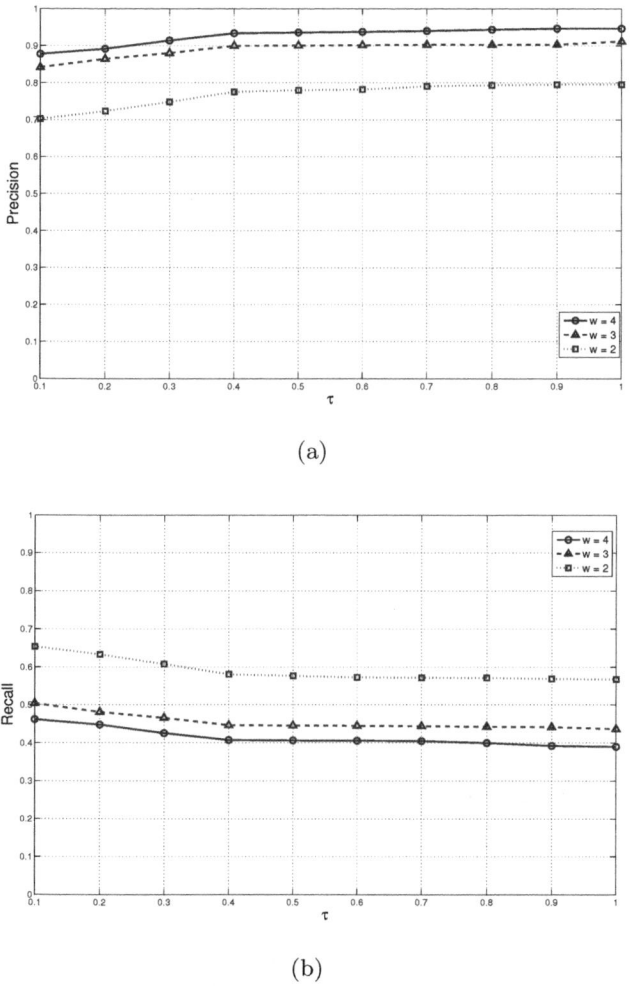

(a)

(b)

Fig. 4. Impact of w on precision and recall measures when $\sigma = 0.1$

varying the threshold τ from 0.1 to 1. Moreover, in order to evaluate the impact of window size w on the quality of predictions, we ranged w from 2 to 4, by considering the predictions done on just one disease unreliable. The results has been obtained by fixing the overall support σ for the frequent patterns to 0.1. Notice that a low support value is necessary for ensuring an adequate length for the mined patterns also in the case of poor cluster homogeneity. As expected, the results in Figure 4(a) clearly reveal that the precision increases as a larger portions of patient medical history, i.e. an increasing number of diseases are used to compute predictions. Conversely, the recall is negatively biased by larger window sizes, as pointed out by Figure 4(b).

After that, for the sake of comparison, we want to show that the overall prediction performances of $CORE^+$ are better than those obtained by the approach $CORE$ [7] . In order to perform a fair comparison, we recur to a well-known metric, the *F-measure* [19], which is the harmonic mean between precision and recall, and it is often used to examine the tradeoff between them:

$$F - measure = \frac{2 * precision * recall}{recall + precision}$$

For this experiment we fixed, for both the approaches, $w = 4$, τ varying from 0.1 to 1, and $\sigma = 0.1$. Figure 5 clearly shows the better overall performances of $CORE^+$ w.r.t. $CORE$. This proves that adding the Markov Models as further layer of prediction, combined with association rules is meaningful.

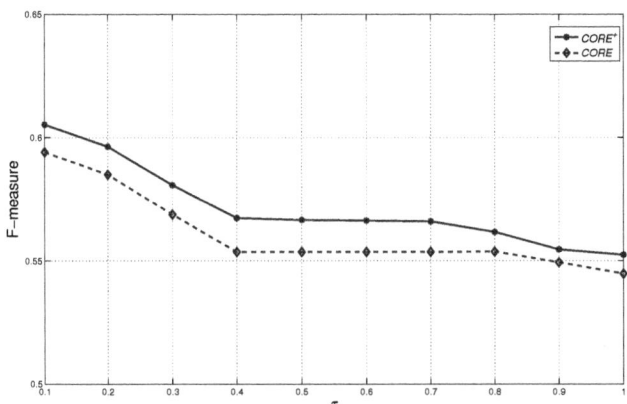

Fig. 5. *F*-measure when $w = 4$, $\tau \in [0.1, 1]$, $\sigma = 0.1$

5   Conclusions

A method based on the combination of clustering, Markov models, and association rules for disease prediction has been presented. The approach uses the past medical history of patients to determine the diseases an individual could incur in the future. As experimental results showed, the combination of more models allows to improve prediction accuracy. The technique can thus be considered a feasible approach to disease prediction. Future works aims to compare our method with other proposals in the literature, and to perform a more extensive evaluation on large medical history data sets.

Acknowledgements. This work has been partially supported by the project *MERIT : MEdical Research in Italy*, funded by MIUR.

References

1. Mooney, R., Strehk, A., Ghosh, J.: Impact of similarity measures on web-page clustering. In: Proc. of AAAI Workshop on AI for Web Search, pp. 58–64 (2000)
2. Agrawal, R., Srikant, R.: Fast algorithms for mining association rules in large databases. In: Proc. of Int. Conf. on Very Large Databases (VLDB 1994), pp. 487–499 (1994)
3. Agrawal, R., Srikant, R.: Mining sequential patterns: Generalizations and performance improvements. In: Apers, P.M.G., Bouzeghoub, M., Gardarin, G. (eds.) EDBT 1996. LNCS, vol. 1057, Springer, Heidelberg (1996)
4. Davis, D.A., Chawla, N.V., Christakis, N.A., Barabási, A.L.: Time to CARE: a collaborative engine for practical disease prediction. Data Mining and Knowledge Discovery 20, 388–415 (2010)
5. Davis, D.A., Chawla, N.V., Blumm, N., Christakis, N.A., Barabási, A.-L.: Predicting individual disease risk based on medical history. In: Proceedings of the ACM International Conference on Information and Knowledge Management (CIKM 2008), pp. 769–778 (2008)
6. Deshpande, M., Karypis, G.: Selective Markov models for predicting web page accesses. ACM Trans. Internet Techn. 4(2), 163–184 (2004)
7. Folino, F., Pizzuti, C.: A comorbidity-based recommendation engine for disease prediction. In: Proc. of 23rd IEEE International Symposium on Computer-Based Medical Systems (CBMS 2010), pp. 6–12 (2010)
8. Folino, F., Pizzuti, C., Ventura, M.: A comorbidity network approach to predict disease risk. In: Khuri, S., Lhotská, L., Pisanti, N. (eds.) ITBAM 2010. LNCS, vol. 6266, pp. 102–109. Springer, Heidelberg (2010)
9. Jaccard, P.: The distribution of the flora of the alpine zone. New Phytologist 11, 37–50 (1912)
10. Jain, A.K., Murty, M.N., Flynn, P.J.: Data clustering: a review. ACM Computing Survey 31(3), 264–323 (1999)
11. Khalil, F., Li, J., Wang, H.: A framework of combining Markov model with association rules for predicting web page accesses. In: Proc. of the Fifth Australasian Data Mining Conference (AusDM 2006), pp. 177–184 (2008)
12. Khalil, F., Li, J., Wang, H.: Integrating recommendation models for improved web page prediction accuracy. In: Proc. of the 31st Australasian Computer Science Conference (ACSC 2008), pp. 91–100 (2008)
13. Khalil, F., Li, J., Wang, H.: An integrated model for next page access prediction. I. J. Knowledge and Web Intelligence 1(1/2), 48–80 (2009)
14. Mongomery, D., Runger, G.: Applied Statistics and Probability for Engineers. John Wiley and Sons Inc., Chichester (2004)

15. Papoulis, A.: Probability, Random Variables, and Stochastic Processes. MacGraw Hill, New York (1991)
16. Pei, J., Han, J., Mortazavi-Asl, B., Pinto, H., Chen, Q., Dayal, U., Hsu, M.: Prefixspan: Mining sequential patterns by prefix-projected growth. In: Proc. of 17th International Conference on Data Engineering (ICDE 2001), pp. 215–224 (2001)
17. Snyderman, R.: Prospective medicine: The next health care transformation. Academic Medicine 78(11), 1079–1084 (2003)
18. Steinhaeuser, K., Chawla, N.V.: A network-based approach to understanding and predicting diseases. Social Computing and Behavioral Modeling (2009)
19. Tan, P., Steinbach, M., Kumar, V.: Introduction to Data Mining. Pearson International Edition, London (2006)

Superiority Real-Time Cardiac Arrhythmias Detection Using Trigger Learning Method

Mohamed Ezzeldin A. Bashir[1], Kwang Sun Ryu[1], Soo Ho Park[1], Dong Gyu Lee[1],
Jang-Whan Bae[2], Ho Sun Shon[1], and Keun Ho Ryu[1,*]

[1] Database/Bioinformatics Laboratory,
Chungbuk National University, Korea
{mohamed,ksryu,soohopark,dglee,shon0621,
khryu}@dblab.chungbuk.ac.kr
[2] College of Medicine, Chungbuk National University,
Cheongju City, South Korea
drcorazon@hanmail.net

Abstract. The Electrocardiogram (ECG) signal uses by Clinicians to extract very useful information about the functional status of the heart, accurate and computationally efficient means of classifying cardiac arrhythmias has been the subject of considerable research efforts in recent years. The contradicting considerations on the unique characteristics of patient's activities and the inherent requirements of real-time heart monitoring pose challenges for practical implementation. That is due to susceptibility to potentially changing morphology not only between different patients or patient cluster, but also within the same patient. As a result, the model constructed using an old training data no longer needs to be adapt with the new concepts. Consequently, developing one classifier model to satisfy all patients in different situation using static training datasets is unsuccessful. Our proposed methodology automatically trains the classifier model by up-to-date training data, so as to be identifying with the new concepts. The performance of the trigger method is evaluated using various approaches. The results demonstrate the effectiveness of our proposed technique, and they suggest that it can be used to enhance the performance of new intelligent assistance diagnosis systems.

Keywords: Electrocardiogram (ECG), Arrhythmias, training dataset, cardiac monitoring, and Classification.

1 Introduction

The Electrocardiogram (ECG) is a series of waves and deflections recording the cardiac's (heart) electrical activity sensed by several electrodes, known as leads. ECG signals generated by sensing the current wave sequence related to each cardiac beat. The P wave employed to represent the atrial depolarization, QRS complex for ventricular depolarization and T wave for ventricular repolarization. Fig. 1 depicts the basics shape of a healthy ECG heartbeat signal.

---

* Corresponding author.

C. Böhm et al. (Eds.): ITBAM 2011, LNCS 6865, pp. 53–65, 2011.

ECG signals are a very important medical instrument that can be utilized by clinicians to extract very useful information about the functional status of the heart. So as to detect heart's arrhythmia, which is the anomalous heartbeat, mapped with a different shape in the ECG signal noticed by deflection on the P, QRS, and T waves, which acquired by some parameters and an enormous finding is produced [1]. Considering the layout procedures of detecting the heart arrhythmias in real time, which begins with extracting the ECG signals, filtering, specifying the features and descriptors, selecting the training datasets, and end with constructing the classifier model to specify the types of arrhythmia in an accurate manner [2].

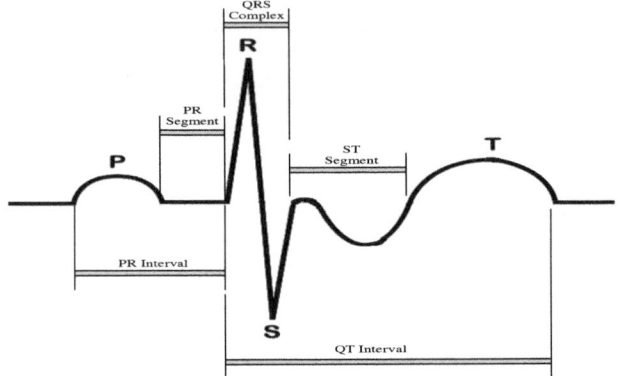

Fig. 1. Shape of a healthy ECG heartbeat signal

There has been a great deal of interest in the systems that provide real-time ECG classification through an intermediary local computer between the sensor and the control center [3]. It's vital for the automated system to accurately detect and classify ECG signals very fast, to provide a useful means for tracing health of the heart in the right time. The effectiveness of such systems is affected by several factors, including the ECG signals, the estimated ECG's features and descriptors, the dataset used for learning purpose and the classification model [4].

Wired ECG monitoring in hospitals are very crucial for saving people's lives. But, this kind of monitoring is inadequate for coronary cardiac disease's patients, who need to be followed up at home, and in open air, as well as for those who need a continued monitoring system to save their life. The morphological descriptors vary from person to person, even within one individual it differs from time to time. Accordingly, building a classifier model with static training data set no longer will be very poor since it will run out of the scope. Consequently, limited quantity and quality of arrhythmias' detection are achieves.

In this paper, we are addressing the challenges for training the classifiers model with updated data to facilities the process of developing real-time cardiac health monitoring systems in details. Furthermore, we present a trigger method as a proposed solution to solve that problem. The performance of the trigger algorithms has been evaluated using various approaches. Consequently, the results demonstrate the effectiveness of our propos method.

In the rest of this paper we will provide a brief background of related work, the description of the trigger methods is follow, then the experimental works, and finally the conclusion.

2   Related Work

In the following subsections we will introduce the most well-known philosophies used to select the training dataset, and incremental learning algorithms utilized to classify the arrhythmias in real time.

2.1   Arrhythmias Detection Training Dataset

Confirming local and global dataset as the main two approaches of training dataset used to learn the classifier model. Global dataset is built from a large database which most automatic ECG analysis research works are referred to this technique such as [4], [5]. Simply, there are training and test dataset with different percentages through which the classifier is going to train using the training dataset to predict the unseen group of data through the test dataset. (Rodriguez, J. et) attempted to drive approach that can build the most accurate model for classifying cardiac arrhythmias based on features extraction: divided the dataset into random groups one training (66%) and another for validation (33%) [6]. One main challenge faced by this technique is the morphologies of the ECG waveforms that widely vary from patient to patient. Accordingly, the classifier learned by specific data related to an identical patient will perform very well when tested with unseen data of that patient often fails when present with ECG waveforms for another patient. To overcome this problem, the literature shows that there is a trend to learn the classifier by training dataset as much as possible. That was the commercial trend introduced by the ECG devices vendors. Though, such approach criticizes in different aspects: first, when using huge amount of ECG records to build a classifier, it will become very complicated in the ways of development, maintaining, and updating. Second, it is difficult to learn the classifier by abnormalities of the ECG during the monitoring process. Hence, there is a possibility to be unable to detect specific arrhythmia when applying that classifier on patients' records. Moreover, it is impossible to introduce all the ECG waveform from all expected patients [7]. Y. H. Hu et al. [8] overcomes this problem through a technique that makes the customization of the training dataset self-organizing, by utilizing the mixture-of-expert (MOE) to realize patient adaptation. This means there is no need to introduce manual distribution of the overall database, the thing that relieves the time and effort consumption. However, such an approach suffers from several pitfalls like: the lack of sensitivity due to the time spending to compare between the two experts (original classifier and the patient specific classifier). Moreover, it is error pruning due to the dependability on different classifiers.

The local learning set is a customized set to a specific patient. In other word, it is a technique focuses on developing a private learning dataset corresponding to each patient [9]. Accordingly, familiarize the classification model with the unique characteristics of each patient. Although this technique looks to alleviate the problem of the learning process, it suffers from a clear problem related to the difficulties to distribute ECG database in relation to time consuming and labor intensive task. Moreover, few patients are accepted to be involved in the development of the ECG processing method. Thus, there are limitations to the advantages provided by such a technique among the expected audience even if it is permissible.

In previous work we suggested a nested ensemble technique to solve the problem of training dataset by manipulating the training dataset for learning the classifier through up-to-date data, and manipulating the ECG features to select the proper adequate set (morphological features) to enrich the accuracy [10]. Although, the results are favorable, the synchronizing of the two components however, is expensive, which affects the detection of the arrhythmias in real time negatively. Moreover, it is quite static to some extent.

2.2 Arrhythmias Classification Methods

A supervised training technique was used to build a model for classifying the ECG data. The classifier model maps the input features to the required output classes, using adjustable parameters specified during the training process. Automated arrhythmias classification using the ECG features (P, QRS, and T) was traditionally performed using supervised and non-supervised methods. Several data mining techniques were used for this intent. One of the most famous techniques used to classify the cardiac arrhythmias is utilizing the decision-tree based on different features [11], [12].

Detecting arrhythmias by applying the pattern recognition methods are very well known. The detection process starts by learning the model with different shapes of ECG parameters during the training session, extracting different statistical parameters of these ECG training dataset, and later using these parameters to classify the unseen ECG during the testing session. Several efforts have been made to apply Artificial Neural Networks (ANNs) for the purpose of heart arrhythmias classification and detecting the cardiac abnormalities. ANNs has compensation of good noise tolerance, in addition to its high efficiency when dealing with non-linear problems [13- 18]. But, this technique has being suffering from many drawbacks such as: the limited numbers of arrhythmias that can be detected due to the restricted number of genuine arrhythmia shapes need to be saved in the memory for a matching purpose. That is, besides the computational time that rises rapidly when increasing the number of arrhythmias aimed to be classified, the things that leads to impracticality in real life. Other methods like support vector machine [19], nearest neighbor [20], rule base classifier [21], and fuzzy adaptive [22] are also introduced in this area. Generally, these methods are general purpose and can be applied in any classification task. The judgment upon such techniques bases on accuracy: the right description of the arrhythmia, effectiveness: the sensitivity to detect the abnormalities in the same time when it took place, efficiency: the speed by which the class of the arrhythmia is going to be specified, and the reliability of the classifier: how far doctors can trust that model to judge future unseen ECG data. These factors are fluctuating from one to another method.

3   Trigger Learning Method

Traditionally, the computation process to detect arrhythmias starts with detecting the ECG signal, filtering and extracting the useful features, training the classifiers, and specifying the type of the rhythm within limited numbers of labels as shown in fig. 2 (left). In the traditional mining approach, errors in early stages such as feature extraction affect the overall performance of the method. Ambiguous output persists and might not be resolved using single learning. Moreover, the dependence on only one learning often makes errors that are apparent from a classifier model.

The trigger method has been suggested to serve the detection of the cardiac arrhythmias on line in very sufficient manner, simply introduced to learn the classifier model by up-to-date training data. Fig. 2 (right) illustrates our idea.

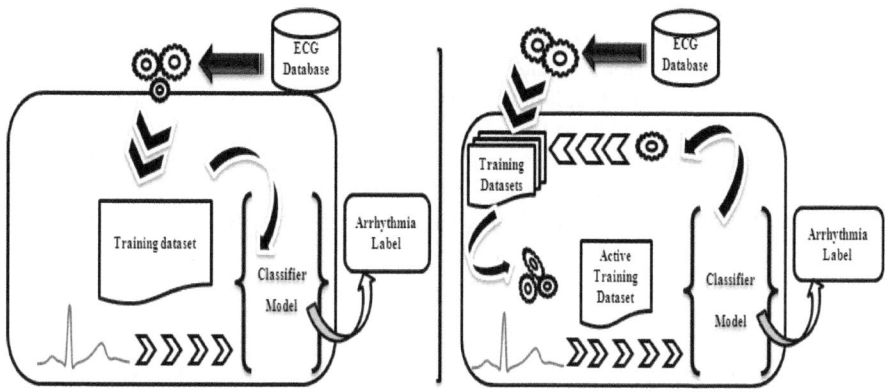

Fig. 2. Traditional training approach and the proposed trigger approach

The trigger technique has four steps as shown in fig. 3. First, an initial learning stage is introduced to learn the classifier by a random set of data without any further consideration. The classifier performance is evaluated (check) and updated (improve) for consistency, and apply the re-movement stage to avoided a combat situation.

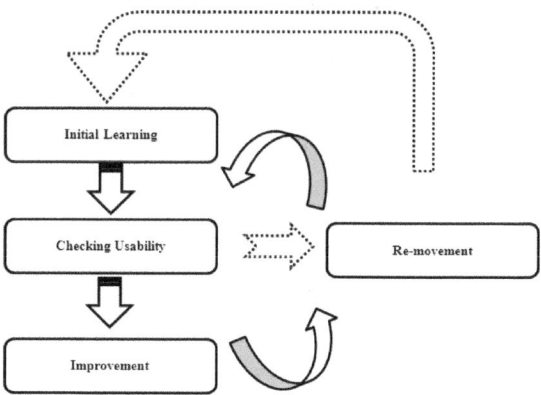

Fig. 3. Learning process flow

3.1 Initial Learning Step

First, we start the learning process by utilizing random group of records (categories), which represent (40%) from the overall dataset without considering any factors or any details for consideration, so as to start on the process of labeling (detecting arrhythmia types). And later on performing (check) and (improve) steps to ensure the correctness of the arrhythmias assignment process, when applying the classifier model to deal with testing data that represent (60%) from the overall dataset.

3.2 Checking Usability Step

After the initial step, the assigned labels are checked in randomly selected patients' records with specific attributes. These processes are conducted based on the overall trusted mark $Trust^M(x)$ which calculated using the local trusted mark $L^M(x)$ that can be measured using label assigned to the specific category with specific vector of features. If the label of the same category (with the same feature sets) is assigned to the target category, then the local trusted mark $L^M(x)$ which represents the local dependability will increase. Local trusted mark $L^M(x)$ calculated with the following formula:

$$l^M(x) = \sum_{i \in 1}^{n} \beta_i(c).C^{label(i)}(i) \tag{1}$$

Where (i) is a range of labels (arrhythmias), $label (i)$ indicates the label assigned to category (c), and $C^{label\ (i)}(i)$ represents the category score $C^S(x)$ labeled by (i). The category score $C^S(x)$ is calculated as follows where features (f) are the feature set related to the category (c):

$$c^s(x) = \sum_{i=1}^{n} c^s features(f) \tag{2}$$

The function $\beta_i(c)$ checks the label assigned to category (c). It returns '+1' if the label (i) is assigned to category (c), otherwise it returns '−1'.

$$\beta_i(c) = \begin{cases} +1 \text{ if label(i)} = c \\ -1 \text{ otherwise} \end{cases} \tag{3}$$

The local trusted marks $L^M(x)$ dependability's are considered on the overall trusted mark $Trust^M(x)$, which is defined using a sigmoid function $Sigmoid(x)$ ($0.5 < Trust^M(x) < 1$):

$$Trust^m(x) = sigmoid \sum_{i=1}^{n} l^m(x) \tag{4}$$

$$sigmoid(x) = \frac{1}{1 + \exp(x)} \tag{5}$$

The overall $Trust^M(x)$ is utilized as a likelihood that indicates the usability of the training set (X) containing specific categories. In case of the $Trust^M(x)$ is greater than some arbitrarily chosen number, (X) is judged to be reliable i.e. active, otherwise (X) is judged to be unreliable i.e. not active. The unreliable (X) is either improved or removed. The overall $Trust^M(x)$ fluctuates continuously in relation to the overall performance of the classifier model and it is ability to detect the right label of the arrhythmia.

3.3   Improvement Step

The checking step end by two judgments, either the current training set is reliable or not, depending on different classes of arrhythmias assigned to categories. Accordingly, the unreliable set needs to be modified by new group of data. This process has two parts first specifying the useless category or categories, and secondly replacing it, or them, with new selected one(s). In the first part the category (c) in the current active training set (X) is removed if the *category score* $C^S(x)$ is less than a threshold δ_{remove}. The remove process will be according to the following formula:

$$\text{if } \{C^S(x) < \delta_{remove}$$
$$\text{then remove} \tag{6}$$

Second the selection of a new category is held randomly depending on a probability $pc'(x)$ that specific category (c) is going to be used in updating the current training set (X). The probability $pc'(x)$ is relative to overall $Trust^M(x)$ that was calculated in Eq. (4)

$$p^{c'} = \frac{trust^{M'}(x)}{\sum_{i \neq c} trust^i(x)} \tag{7}$$

The substitute category, which is selected when utilizing the probability $p^{C'}(x)$, is newly assigned to the active training group of data (active xs). Then, the process returns to the loop of the check and improvement steps.

The replacement of the impractical category could be executed several times during the loop of the check and update steps. Categories that are removed from the current active training set (X) have a chance to be selected in the subsequent update step for reactivation, which means all categories have the possibility of being assigned, regardless of the remove process.

3.4   Re-movement Step

The improvement step is very active when there is a limited number of bad labeling using the current training set (X), while it is useless when there are multiple defects among the categories, the thing that requires inherited improvement process consequence. It is very expensive in terms of time, which affects negatively the performance of the classifier model to label different types of arrhythmias. Therefore, the re-movement step is introduced to deal with this problem.

All categories in *(X)* are removed i.e. the active training set is removed if it is defect score $D^S(x)$ is greater than a threshold θ_{remove}. The remove process will be according to the following formula:

$$\text{if } \{D^S(x) > \theta_{remove}$$
$$\text{then remove} \tag{8}$$

In this case, the initial learning step will restart again with the same procedures. But the selection of the categories must be with a new group of categories (not randomly), which can be achieved using equation (7). With overall 40% for training and 60% for testing that never affects neither by improvement step or re-movement step.

4   Experimental Environment

We used a database generated at the University of California, Irvine [23]. It was obtained from Waikato Environment for Knowledge Analysis (WEKA), containing 279 attributes and 452 instances [24]. The classes from 01 to 15 were distributed to describe normal rhythm, Ischemic changes (Coronary Artery Disease), Old Anterior Myocardial Infarction, Old Inferior Myocardial Infarction, Sinus tachycardy, Sinus bradycardy, Ventricular Premature Contraction (PVC), Supraventricular Premature Contraction, Left bundle branch block, Right bundle branch block, degree AtrioVentricular block, degree AV block, degree AV block, Left ventricule hypertrophy, Atrial Fibrillation or Flutter, and Others types of arrhythmias Respectively.

The experiments were conducted in WEKA 3.6.1 environment. Our experiments were carried out by a PC with an Intel Core processor (T M) 2 DUO, speed to 2.40 GHz. And RAM 2.00 GB. The parameters were set as follows: in equation (6) δ_{remove} = 1.0, and in equation (8) θ_{remove} = 5.0.

5   Results

5.1   Arrhythmias Detection

First of all, we proved the necessity of including the P and T waves in conjunction with the QRS complex to evaluate arrhythmias in the right way. We measure the performance of five different algorithms the OneR, J48, Naïve Bayes, Dagging, and Bagging according to the parameter(s) used to classify the arrhythmias. Table 1 summarizes the results obtained by each algorithm.

Table 1. Accuracy of different algorithms according ECG's parameters

parameter	OneR (%)	J48 (%)	Naïve Bayes (%)	Dagging (%)	Bagging (%)
QRS only	60.4	91.2	76.5	63.5	81.0
QRS + P	60.4	91.4	77	62.4	81.6
QRS + T	61.3	91.2	76.7	63.0	82.3
QRS + P + T	61.1	92.3	77.7	64.2	83.0

Fig. 4 shows the value added by the trigger learning technique to the all algorithms that were mentioned in the previous experiment, we choose the best case when including all features related to the QRS, P, and T waves. We spot-light that by comparing their accuracy's performance before and after using the trigger learning method.

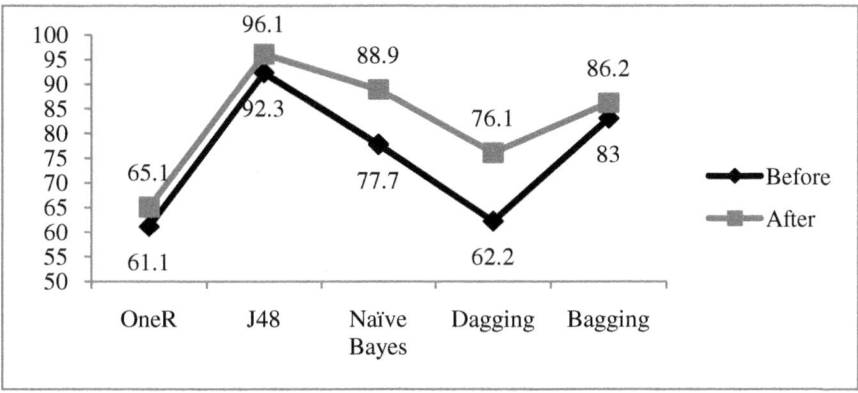

Fig. 4. Improvement addressed by the trigger method

It is clear proof that, the proposed trigger learning method improves the process of detecting the different types of arrhythmia efficiently. Such improvement is noticeable along all the algorithms with different weights due to their mechanisms. Specifically, there were 4, 3.8, 11.2, 13.9, and 3.2 upgrading points were detected in the performance of OneR, J48, Naïve Bayes, Dagging, and Bagging respectively. In general, it is a valuable achievement in the arrhythmias detection problem area.

It is also remarkable to compare the result of our scheme for ECG classification with that of other methods presented in the literature. The patient adaptive model (PAM) [8], principle component with independent component analysis (PCICA) [25], and Gaussian mixture base classifier model (GMM) [26], are selected for this comparison. Table 2 shows the results.

Table 2. Accuracy comparison with other methods

Method	Number of arrhythmias	Accuracy (%)
PAM	4	94.0
PCICA	5	85.0
GMM	2	94.3
trigger method with J48	15	96.1

Among the three methods, the J48 when enhanced with the proposed trigger learning method outperforms the others with an impressive accuracy of 96.1% to discriminate 15 ECG beat types.

5.2   Atrial Fibrillation (AF) Detection

Secondly we measure the performance of the trigger learning method to detect only one arrhythmia. Accordingly, we selected the atrial fibrillation (AF), which is the most frequently occurring cardiac arrhythmia. It is the major cause of morbidity in populations over the age of 75 [27]. AF causes the heart to beat irregularly, leading to inefficient pumping of blood and changing the blood flow dynamics. AF arrhythmia, nature affects all the parameters by modifying their shapes and intervals as it shown in Figure 5.

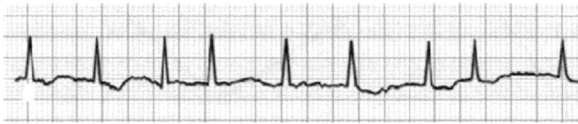

Fig. 5. AF Arrhythmia Shape

We generate a subset of the same database [23], which contains only upon Atrial Fibrillation arrhythmia and normal rhythm. We duplicate the AF and generate a total of 249 instances, including 21 cases of Atrial Fibrillation with the rest a normal rhythm. And then, trigger learning method was applied using J48 algorithm.

For a detailed performance analysis of the trigger learning, the sensitivity, specificity, and accuracy were obtained. The classification performance is generally presented by a confusion matrix where the TP, TN, FP, and FN stand for true positive, true negative, false positive, and false negative, respectively.

Accordingly, we evaluated: the accuracy, expressed in percentage of the division of the sum of correctly detected AF (TP+TN) by the sum of all parameters (TP+TN+FP+ FN), provides a measure of the precision of the algorithm; sensitivity, expressed in percentage of the division of all true AF (TP) by the sum of TP + FN, provides a measure of the capacity of the technique to detect AF; specificity, expressed in percentage of the division of all non- Atrial Fibrillation AF (TN) by the sum of TN + FP, provides a measure of the capacity of the technique to confirm the non-presence of AF episodes in the ECG. Table 3 shows the parameters figures obtained when applying the trigger learning and J48 to detect the AF. The results imply that it has good predictive abilities and generalization performance.

Table 3. Trigger learning performance with different ECG's parameters

parameters	TP	FN	FP	TN	sensitivity	specificity	accuracy
QRS only	17	3	4	225	85.5%	98.3%	97.2%
QRS + P	18	3	3	225	85.7%	98.7%	97.5%
QRS +P +T	19	1	1	228	95.0%	99.6%	99.2%

Based on the results, the trigger learning technique has provided slightly better performance when applying with the QRS complex and the P waves than when utilizing only the QRS complex. In contrast, when introducing the QRS complex, P, and T waves we got an outstanding performance.

Several researchers have addressed the AF arrhythmia detection problem using the ECG signals directly or by analyzing the heart rate variability signal [27–30]. Generally, all these techniques are utilizing either QRS complex mainly the R wave, or the P waves. The literature never shows the employment of other ECG parameters and their intervals to detect AF arrhythmia. Table 4 summarizes the testing results obtained by different methods. It can be observed from this table that the models derived using trigger learning and QRS complex, P, and T waves provide a better accuracy than those obtained by other methods that are reported in the literature.

Table 4. Comparative results of different AF detection methods

Author	Database	Method	sensitivity	Specificity	Accuracy
Christov et al. [28]	ECG	P wave	95.7%	-	98.8%
Logan and Healey [29]	HRV	RR irregularity	96%	89%	-
Rodriguez and Silveira [30]	HRV	RR interval	91.4%	-	-
Fukunami et al. [31]	ECG	P wave	91%	76%	-
Trigger methods	ECG	QRS +P+T	95%	99.6%	99.2%

6 Conclusions

Detecting the heart arrhythmias through ECG monitoring is mature research achievement. Wired ECG monitoring in hospitals are very crucial for saving people's lives. However, this kind of monitoring is insufficient for coronary cardiac disease's patients, who need continued follow ups.

The contradicting considerations on the unique characteristics of patient's activities and the inherent requirements of real-time heart monitoring pose challenges for practical implementation. The outsized morphological disparity of the ECG is seen not only between different patients or a patient cluster but also within the same patient and as a result, the model constructed using the old training data no longer needs to be adjusted in order to identify with the new concepts. In view of that, developing one classifier model to satisfy all patients in different situations using static training datasets is unsuccessful.

In this paper, we presented a trigger method as a proposed solution to solve this problem. The performance of the trigger algorithms has been evaluated using various approaches. Furthermore, the results demonstrate the effectiveness of our proposed method. In the future, we plan to improve the trigger method much more to detail with noise since its sensitivity is very high.

Acknowledgment. This work was supported by the grant of the Korean Ministry of Education, Science and Technology (The Regional Core Research Program / Chungbuk BIT Research-Oriented University Consortium), and the Basic Science Research Program through the National Research Foundation of Korea (NRF) funded by the Ministry of Education, Science and Technology (NRF No. 2011-0001044).

References

1. Dale Dubin, M.: Rapid interpretation of EKG's, 6th edn. cover publishing co. (2000)
2. Rajendra, U., Sankaranarayanan, M., Nayak, J., Xiang, C., Tamura, T.: Automatic Identification of cardiac health using modeling techniques: a comparative study. Inf. Science 178, 4571–4582 (2008)
3. Rodrigues, J., Goni, A., Illarramendi, A.: Real time classification of ECG on a PDA. IEEE Trans. on IT in B. Med., 23–33 (2005)
4. Bashir, M.E.A., Akasha, M., Lee, D.G., Yi, M., Ryu, K.H., Bae, E.J., Cho, M., Yoo, C.: Highlighting the Current Issues with Pride Suggestions for Improving the Performance of Real Time Cardiac Health Monitoring. In: Bringas, P.G., Hameurlain, A., Quirchmayr, G. (eds.) DEXA 2010. LNCS, vol. 6262, Springer, Heidelberg (2010)
5. Bortolan, G., Jekova, I., Christov, I.: Comparison of four methods for premature ventricular contractions and normal beats clustering. Com. Card. 30, 921–924 (2005)
6. Rajendra, U., Subbann, P., Iyengar, S., Raod, A., Dua, S.: Classification of heart rate data using artificial neural network and fuzzy equivalence relation. Pattern Recognition, 61–68 (2003)
7. Clifford, G., Azuaje, F., McSharrg, P.: Advanced methods and tools for ECG data analysis. Artech house, Boston (2006)
8. Hu, Y.H., Palreddy, S., Tompkins, W.J.: Patient adaptable ECG beat classification using mixture of experts. In: Neural Network for Signal Processing V, pp. 463–495. IEEE Press, Piscataway (1995)
9. Palreddy, H., Tompkins, W.: A patient-adaptable ECG beat classifier using a mixture of experts approach. Trans. On B. Med. Eng. 44, 891–900 (1997)
10. Bashir, M.E.A., Akasha, M., Lee, D.G., Yi, M., Ryu, K.H., Bae, E.J., Cho, M., Yoo, C.: Nested Ensemble Technique for Excellence Real Time Cardiac Health Monitoring. In: BioComp., lasvegas, USA (2010)
11. Abenstein, J.P.: Algorithms for real time ambulatory ECG monitoring. Biomed. Sci. Instrum. 14, 73–79 (1978)
12. Drazen, E.L., Garneau, E.F.: Use of computer-assisted ECG interpretation in the United States. In: Proc. Computers in Cardiology (1979)
13. de Bie, J.: P-wave trending: A valuable tool for documenting supraventricular arrhythmias and AV conduction disturbances, pp. 511–514. IEEE, Los Alamitos (1991)
14. Bortolan, G., Degani, R., Willems, J.L.: ECG classification with neural networks and cluster analysis. In: Proc. Computers in Cardiology, pp. 177–180 (1991)
15. Chang, J.C.: Applying artificial neural network for ECG QRS Detection.: Master thesis, Univ. of Wisconsin—Madison (1993)
16. Hu, Y.H., Tompkins, W.J., Urrusti, J.L., Alfonso, V.X.: Applications of artificial neural networks for ECG signal detection and classification. Electrocardio (1994)
17. Suzuki, Y., Ono, K.: Personal computer system for ECG ST-segment recognition based on neural networks. Med. Biol. Eng. Computing, 2–8 (1992)

18. Watrous, R., Towell, G.: A patient-adaptive neural network ECG patient monitoring algorithm. Comput. Cardiol. (1995)
19. Yang, T., Devine, B., Macfarlane, P.: Artificial neural networks for the diagnosis of atrial fibrillation. Med. Biol. Eng. Comp. (1994)
20. Kampouraki, A., Manis, G., Nikou, C.: Heartbeat time series classification with support vector machines. Eng. in Med. and Bio. Sc., 512–518 (2009)
21. Christov, I., Bortolan, G.: Ranking of pattern recognition parameters for premature ventricular contractions classification by neural networks. Phys. Measure, 1281–1290 (2004)
22. Birman, K.: Rule-Based Learning for More Accurate ECG Analysis. Tran. On Pattern Analysis and Mach. Int., 369–380 (1982)
23. UCI Machine Learning Repository,
    http://www.ics.uci.edu/~mlearn/MLRepository.html
24. WEKA web site, http://www.cs.waikato.ac.nz/~ml/weka/index.html
25. Yu, S.N., Chou, K.T.: Integration of independent component analysis and neural networks for ECG beat classification. Expert Systems with Applications 34, 2841–2846 (2008)
26. Martisa, R.J., Chakrabortya, C., Ray, A.K.: A two-stage mechanism for registration and classification of ECG using Gaussian mixture model. Pattern Recognition 42, 2979–2988 (2009)
27. Wheeldon, N.M.: Atrial fibrillation and anticoagulant therapy. Euro. Heart J. 16, 302–312 (1995)
28. Christov, I., Bortolan, G., Daskalov, I.: Sequential analysis for automatic detection of atrial fibrillation and flutter. Comput. in Cardiol., 293–296 (2001)
29. Logan, B., Healey, J.: Robust detection of atrial fibrillation for a long term telemonitoring system. Computer Cardiol. 32, 619–622 (2005)
30. Rodriguez, C.A.R., Silveira, M.A.H.: Multi-thread implementation of a fuzzy neural network for automatic ECG arrhythmia detection. Comput. Cardiol. 28, 297–300 (2001)
31. Fukunami, M., Yamada, T., Ohmori, M., Kumagai, K., Umemoto, K., Sakai, A., Kondoh, N., Minamino, T., Hoki, N.: Detection of patients at risk for paroxysmal atrial fibrillation during sinus rhythm by P wave-triggered signal-averaged electrocardiogram. Circulation 83, 162–169 (1991)

Monitoring of Physiological Signs
Using Telemonitoring System

Jan Havlík[1], Jan Dvořák[1], Jakub Parák[1], and Lenka Lhotská[2]

[1] Department of Circuit Theory, Faculty of Electrical Engineering
Czech Technical University in Prague, Technická 2, CZ-16627 Prague 6
xhavlikj@fel.cvut.cz
[2] Department of Cybernetics, Faculty of Electrical Engineering
Czech Technical University in Prague, Karlovo nám. 13, CZ-12135 Prague 2

Abstract. The paper presents telemonitoring system for distant monitoring of physiological signs. The system is designed as modular system content input modules for monitoring various vital signs like ECG, NIBP and oxygen saturation, control module and telecommunication modules for streaming data using various wireless technologies like Bluetooth, GSM and WiFi. The input and telecommunication modules are interchangeable. The system is able to preprocess acquired signals using filtration, parameterization etc. It allows stream both raw data and only aggregated data. The monitoring part is supplemented by PC based part for storing data in database, prospective processing of data and sharing data with other systems.

Keywords: telemonitoring, vital signs monitoring, smart home, assistive technology.

1 Introduction

The paper presents telemonitoring system for distant monitoring of physiological signs. The telemonitoring is one of the up-to-date techniques in the field of assisted technologies and smart homes for distant monitoring of vital signs. This technique is able to monitor patients in their own environment continuously without significant limitations and to provide information about the health of persons under monitoring [1–4].

2 Design and Realization

The presented system is designed as a system for distant monitoring of physiological signs like electrocardiogram (ECG), peripheral blood oxygen saturation (SpO$_2$) and non-invasive blood pressure (NIBP). The system is modular and consists of three independent parts – input modules, control unit and telecommunication modules. The main task of the system is to sense some vital sign, to process acquired signals and to communicate them to PC based system using any type of standardized wireless technologies such as Bluetooth, WiFi or

C. Böhm et al. (Eds.): ITBAM 2011, LNCS 6865, pp. 66–67, 2011.

GSM. The input modules transduce measured biosignal to appropriate electrical value (digital data or voltage frequently). The control unit serves several tasks simultaneously. The important ones are to acquire input signals and to convert them to digital data (of course for analog inputs only), to process these signals and/or to parameterize them, to prepare data packets according to defined communication protocol and to send the data. The control unit has to provide the user interface of whole system also. The communication modules support the transmission of signals between control unit and PC based system by any type of wireless techniques like Bluetooth, GSM, WiFi etc. The interfaces between modules are strictly defined and the modules are reciprocally interchangeable. It means it is possible to choose measured signal and the type of connection and to assemble user required system in easy and quick way.

The monitoring part is supplemented by PC based application for storing data in database, prospective processing of data and sharing data with other systems.

3   Conclusion

The modular system for telemonitoring of physiological signs has been described shortly in this paper. The system is user modifiable for wide range of use. It could be a fundament for complex equipment of smart home and could be a base of more sophisticated system in the field of assistive technologies.

Acknowledgement. This work has been supported by the grant No. F3a 2122/2011 presented by University Development Foundation and by the research program No. MSM 6840770012 of the Czech Technical University in Prague (sponsored by the Ministry of Education, Youth and Sports of the Czech Republic).

References

1. Celler, B.G., et al.: Remote monitoring of health status of the elderly at home. A multidisciplinary project on aging at the University of New South Wales. International Journal of Bio-medical Computing 40(2), 147–155 (1995)
2. Chan, M., Esteve, D., Escriba, C., Campo, E.: A review of smart homes – Present state and future challenges. Computer Methods and Programs in Biomedicine 91(1), 55–81 (2008)
3. Costin, H., et al.: Complex telemonitoring of patients and elderly people for telemedical and homecare services. In: New Aspects of Biomedical Electronics and Biomedical Informatics, pp. 183–187 (2008)
4. Dang, S., Golden, A.G., Cheung, H.S., Roos, B.A.: Telemedicine Applications in Geriatrics. Brocklehurst's Textbook of Geriatric Medicine and Gerontology, pp. 1064–1069. W.B. Saunders, Philadelphia (2010)

SciProv: An Architecture for Semantic Query in Provenance Metadata on e-Science Context

Wander Gaspar, Regina Braga, and Fernanda Campos

Computational Modeling Program, Federal University of Juiz de Fora, Brazil
wandergaspar@gmail.com, {regina,fernanda}@ufjf.edu.br

Abstract. This article describes *SciProv*, an architecture that aims to interact with *Scientific Workflow Management Systems* in order to capture and manipulate provenance metadata. For this purpose, *SciProv* adopts an approach based on an abstract model for representing the lineage. This model, called *Open Provenance Model* (OPM), allows that *SciProv* can set up a homogeneous and interoperable infrastructure for handling provenance metadata. As a result, *SciProv* is able to provide a framework for query metadata provenance generated in an e-Science scenario. Moreover, the architecture uses semantic web technology in order to process provenance queries. In this context, using ontologies and inference engines, *SciProv* can make inferences about lineage and, based on these inferences, obtain important results based on extraction of information beyond those that are registered explicitly from the data managed.

Keywords: e-Science, Metadata Provenance, Scientific Workflow, Semantic Web, and Open Provenance Model.

1 Introduction

Since last decades, the process of design, execution and analysis of scientific experiments has increasingly relied on the use of computational tools. Computing became a great tool to Science and has led to an information explosion in many different fields [6].

However, benefits from such tools entail new and complex challenges in scientific research scenario. An important aspect refers to reuse of knowledge. Scientific experiments computationally processed are subject to frequent updates, either by more refined view of researchers or considering changes in components and procedures. Additionally, databases can be updated periodically, discarding previous results. Therefore, knowledge reuse can mean saving time and resources. Another relevant aspect refers to data analysis, which may have little use if scientists are unable to judge their suitability to experiment under study and identify the origin of results. In a scenario of scientific research, part of the significance of obtained data is due to the understanding of generative process.

Such considerations lead to data provenance, which can be described as: data provenance — also called lineage, genealogy or pedigree — is a description of

C. Böhm et al. (Eds.): ITBAM 2011, LNCS 6865, pp. 68–81, 2011.
© Springer-Verlag Berlin Heidelberg 2011

the origins of a data item and also the process by which it was produced. Data provenance helps to form a vision of quality and validity about the information produced in the context of a scientific experiment modeled by computer [3].

Moreover, a growing amount of digital objects is designated for long term preservation. Specialized approaches, models and technologies are needed to guarantee the long-term understandability of the preserved data. Maintaining the authenticity and provenance of the preserved objects for the long term is of great importance, since users must be confident that the objects in the changed environment are authentic [8].

Currently, it is possible to deal with data provenance in the context of workflows modeled in *Scientific Workflow Management Systems* (SWfMS) such as Vistrails [11], Kepler [15], and Taverna [24]. But how to trace the origin of data in a heterogeneous environment? It is necessary to have a system capable of providing interoperability between generated data provenance.

To obtain the benefits derived from data provenance, it is essential that information related to computer simulation that represents the scientific experiment can be captured, modeled and persisted in a suitable form for further use. In this context, it is noted that management of data provenance is an open question that has received significant interest from the scientific community [19] [22] [2] [9].

Among problems studied, it is important to mention the lack of agreement on the scope and how to model data provenance [16]. Considering this issue, this work is based on a model named *Open Provenance Model* (OPM), which aims to define a generic and comprehensive representation of data provenance [18]. OPM intends to become a standard for dealing with data provenance and it is capable of providing interoperability between existing SWfMS.

2   Design Considerations for a Metadata Provenance Approach

Design considerations for a metadata provenance approach refers to proposed solutions for issues such as collecting, managing and querying data lineage. There are papers in the scientific literature that advocate for an approach that is independent of any SWfMS [22] [14]. This strategy offers an alternative to management of data provenance directly from SWfMS.

OPM model presents an abstract, generic, and comprehensive representation to address the problem of metadata provenance. OPM represents provenance through a causal dependency graph, which allows recording the execution history of a given workflow. An OPM graph can be defined as a directed acyclic graph which nodes represent entities considered by the model — artifacts, processes and agents — and which edges represent the causal dependencies between these entities. To facilitate the understanding and promote a shared vision, OPM model proposes a graphical notation for a provenance graph, where artifacts are represented by ellipses, the processes are represented by rectangles and the agents are represented by octagons. Figure 1 shows an OPM graph where edges are identified relating to causal dependencies between entities of the model.

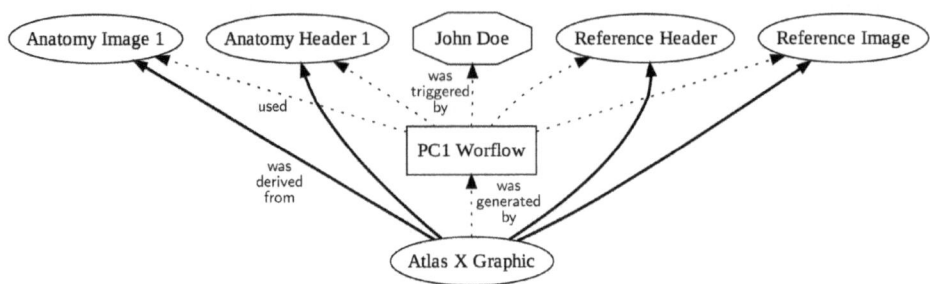

Fig. 1. OPM provenance graph [17]

A capture mechanism is responsible for metadata provenance collecting. Such a mechanism should have access to software artifacts, generated data, and annotations. Capturing mechanisms can act at workflow level, process level, or operating system (OS) level [5]. The former are integrated into SWfMS. Process-based mechanisms require that each workflow activity makes data provenance available. Finally, OS-based mechanisms rely on existing functionality to generate data provenance.

In an approach for collecting and managing metadata provenance independently, it becomes necessary to adopt an instrumentation mechanism, whose goal is to collect lineage data from different SWfMS [22] [19]. In a provenance capture engine that acts on the process level, each workflow component must be instrumented so that data collection can be accomplished. The workflow instrumentation allows provenance capture at various granularity levels. Each process can be instrumented several times, where each adaptation mechanism is built to capture the provenance in different granularity levels as described later.

Several storage solutions have been proposed for metadata provenance using relational DBMS, XML representations or semantic web technology. These solutions are influenced by non-functional requirements associated with project development, including ease of use and restrictions imposed by the execution environment [9]. According to [1], use of a DBMS is the most appropriate alternative to data provenance storage. The choice is based on reliability provided by DBMS and the offer of sophisticated features for data querying. SWfMS such as Redux [7] and Taverna [20] use storage mechanisms based only on RDBMS. Vistrails presents a hybrid mechanism for storing data provenance, which uses both a relational database as well as semi-structured files in XML [9].

An effective query engine is a necessary component of metadata provenance model in the context of scientific experiments. Depending on the complexity of the workflow, the overhead related to query data source may be relevant. [2] propose a strategy that uses abstractions through the creation of user views. According to this model, user indicates which modules are relevant for capturing data provenance and the system processes information according to indicated preferences. A common feature across many of the approaches to querying provenance is that their solutions are closely tied to storage models used. Thus, query languages like SQL, Prolog, or SPARQL are adopted depending on the storage solution used [6]. Although these languages are suitable for querying databases, they were not designed specifically to query provenance data. As a result, the syntax for formulating queries is difficult and complex even when writing simple queries.

Some provenance models employ technologies coming from semantic web for data representation and query. Languages like RDF and OWL provide a more natural approach to metadata provenance and have the potential to simplify interoperability among different models [12]. Information stored in a relational approach can be converted into semi-structured data — for example, in RDF triples — from the specification of an ontology that represents the provenance model adopted. Mechanisms for serializing RDF triples in XML format are particularly interesting in this context.

There are some studies on using metadata provenance in scientific workflows. Both Simmham *et al.* [22] and Marinho *et al.* [16] propose to capture provenance in the context of workflows in a computational grid. The capture of metadata is done through the invocation of web services. However, none of these authors adopts OPM or uses web semantics technology. Furthermore, Marinho *et al.* uses Prolog for the persistence of metadata provenance. Groth *et al.* [13] propose a provenance model based on the concept of Virtual Data System, where part of the data is not persisted. However, this strategy has as its focus — and appears to be efficient — for computational simulations where the volume of information processed is very high.

This paper presents a architecture for query metadata provenance generated in an e-Science scenario. Moreover, the proposed architecture uses semantic web technology in order to process metadata provenance queries. This is the main contribution of this new proposal when compared to previous approaches. Thus, using ontologies and inference engines, *SciProv* can make inferences about lineage and, based on these inferences, obtain important results based on extraction of information beyond those that are explicitly registered from managed data.

Research on metadata provenance is moving fast. New applications fields have emerged like workflow management, data integration, and development tools for a better exploitation of collected data. Thus, we have the *Science Collaboratories* (collaborative scientific laboratories), that consider large-scale deployment of data provenance, and also techniques and tools designed for the manipulation of produced historical information [21]. It is hoped that provenance exploration in a collaborative environment can lead scientists to learn from available examples, capable of accelerating ongoing researches [10].

3   Scientific Workflow Provenance System

The objective of this work is to specify an architecture — called *Scientific Workflow Provenance System* (SciProv) — for data provenance collection and management in the context of scientific experiments processed through computer simulations in collaborative research environments, geographically dispersed and interconnected through a computational grid. *SciProv* should provide an interoperability layer that can interact with SWfMS in order to capture the provenance information generated from scientific workflows.

SciProv is based on OPM abstract model whose vision is to support provenance inter-operability in a precise, technology-agnostic manner. OPM requirements include digital representation of provenance for any "thing', whether produced by computer systems or not [17]. However it is not the purpose of OPM to specify internal representations that systems have to adopt to store or manipulate provenance internally [17]. Although it does not specify protocols to store and query provenance information in provenance repositories, OPM encourage the development of concrete bindings in a variety of languages and the proposal of domain specializations (web, biology, workflows).

Thus *SciProv* represents a specialization in the scientific workflows domain in a manner consistent with the OPM model expectations.

SciProv is based on the following requirements:

– Independence of flow control mechanisms and data formats implemented in any existing SWfMS.
– Applicability targeted but not restricted to scientific experiments, including workflows executed in heterogeneous environments and dispersed in computational grids.
– Use of semantic web technologies — RDF, OWL, ontology, inference machines, SPARQL — to represent and query metadata provenance.
– Possibility to adjust the mechanism to provide tools for monitoring the impact of provenance metadata capture and persistence in implementing the scientific experiment.

3.1 SciProv Architecture

A representation of the proposed architecture considering a typical application scenario is presented in Figure 2. In a collaborative research environment interconnected through a computational grid, a scientific workflow can consist of several components, where outputs of each module corresponds, in most cases, to inputs of the next module.

Researchers can model scientific workflows from distinct SWfMS and whose implementation requires access to heterogeneous repositories (relational data, semistructured, etc.) and distributed in a computational grid. This scenario requires mechanisms that contribute to greater interoperability between data generated and processes orchestrated under collaborative research projects.

The *SciProv* initial responsibility is to provide an adaptive engine called instrumentation for each component that makes up the scientific experiment. This mechanism is implemented using web service technology and configured for each component of the workflow which provenance must be collected and persisted. Instrumentation engine is intended to capture data generated during workflow execution and send them to a centralized relational database. It should be noted

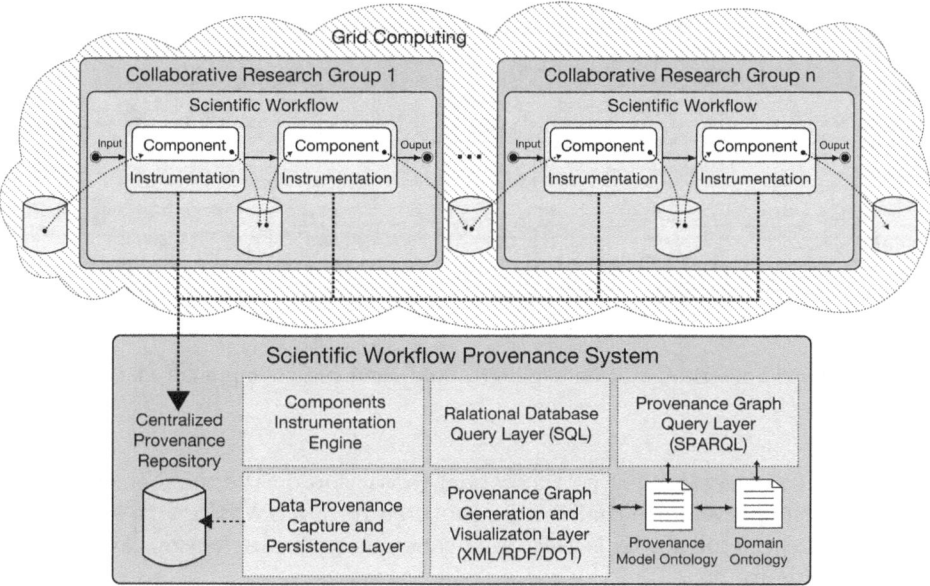

Fig. 2. Scientific Workflow Provenance System architecture

that the process of instrumentation requires a degradation in performance of workflows. Section 4 presents the results of performance measurements and Section 5 presents considerations on this issue.

It can be observed from Figure 3a that *SciProv* web service can be invoked more than once for each workflow component. In a typical scenario, a first web service instance is configured to collect input data (OPM artifacts). A second instance captures information relating to component itself (OPM process) such as start and end execution time, and OPM causal dependencies related to artifacts. A third instance capture the data (artifacts) generated at the component output.

As a result of instrumentation, the original component is replaced by a composite component in the scientific workflow user's view, as shown in Figure 3a. This model allows that scientists can disregard all the complexity related to instrumentation process built into the experiment. It is important to emphasize that the SWfMS that will be used for the experiment must support the use of composite component considering that SciProv instrumentation needs this capability. Kepler, Taverna, and Vistrails provide features that implement the concept.

Considering an approach for collecting provenance at the component level, it opens up the possibility of control over data granularity to be captured. In scientific research, may be important to analyze information and develop queries to data lineage at various levels of detail. The OPM model allows the handling of provenance metadata that comes in many different levels, where each level is associated with a distinct *account*. This concept is fully adopted by *SciProv*. It

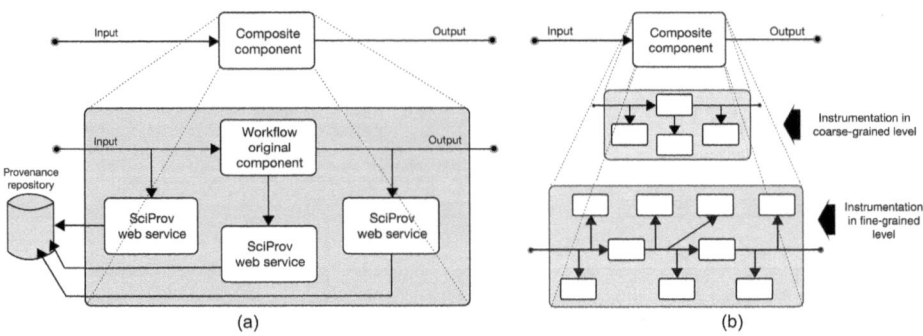

Fig. 3. *SciProv* instrumentation engine (a) and composite component at two levels of abstraction (b)

can be stated that there is no theoretical restriction on the number of levels of detail for the provenance that can be captured from a SWfMS with support for composite components. This feature is coherent with abstraction layers model shown in Figure 3b.

SciProv uses a solution based on relational model to provide the persistence of metadata provenance. Although the implemented prototype adopts a specific RDBMS, the choice of computing infrastructure to support the relational model can be changed without implications or restrictions within the architecture. Because of technical implications for adoption of a distributed storage model, initially *SciProv* prototype provides support for a centralized model of metadata provenance persistence. In this initial approach, it uses a single coordinator node for lineage storage. This solution shows benefits with regard to persisted provenance metadata management and query. However, in scientific experiments with large volume of information handled, data traffic on a computational grid can impact the workflow performance.

One of the most relevant architecture features is to provide an infrastructure based on semantic web technology to process the metadata provenance. The purpose of this approach is to extract information beyond those that are explicitly recorded in the relational database. *SciProv* adopts the approach proposed by *Open Provenance Model* project, which enables that provenance metadata captured from heterogeneous systems can be collected consistently and queried in an integrated manner [17].

The provenance graph generation and visualization layer (Figure 1) has several interrelated tasks. From relational database, metadata persisted according to OPM model is processed in order to obtain a representation of the provenance graph corresponding to the workflow execution. From in-memory provenance graph, *SciProv* provides resources for binding data lineage in semantic web formats.

This layer also builds a graphical representation that allows data lineage visualization in a diagram showing the provenance abstract model entities and causal dependencies. Although this feature is not critical to the architecture, it is interesting to provide a lineage intuitive view in a scientific experiment context. A graphical representation facilitates the understanding of relationships between OPM entities —artifacts, processes, and agents— and causal dependencies that exist in a provenance graph.

One of the distinguishing *SciProv* features consists in provinding provenance graph into RDF format to allow metadata lineage query in SPARQL. According to Figure 2, one can observe that graph provenance binding in RDF format is associated with OPM ontology model and also with domain ontology concerning the scientific experiment.

Finally, the query graph provenance layer aims to provide a mechanism and an interface for user to formulate queries in SPARQL. *SciProv* architecture provides support to ontologies described in semantic web formats as OWL-DL. This feature gives the possibility of processing queries from reasoners able to make inferences about these knowledge bases and get interesting results when extracting information beyond those that are expressed directly on provenance graph. It is important to note that query semantics require not only a set of RDF collected and serialized assertions but also OPM model and domain knowledge representations.

Another important feature promoted by *SciProv* and related to lineage metadata semantic processing is the proposition of an OPM specialization capable of amplifying semantic processing power in scientific workflows domain. This resource aims to enhance reasoners ability to infer about provenance metadata collected in a e-Science scenario (Figure 4).

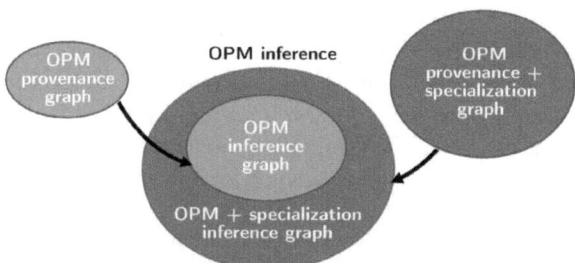

Fig. 4. Richer semantics queries from an OPM model specialization

One of the key component of provenance is user-defined information. This includes documentation that cannot be automatically captured but records important decisions and notes. This data is often captured in the form of annotations [6]. OPM model allows you to attach notes to entities and causal dependencies through a specific framework for this purpose. We can envision the use of annotations as a mechanism to provide greater interoperability between provenance systems and to expand the semantic content for a specific domain studied. *SciProv* uses OPM annotations framework to formulate semantic queries to provenance with results even more rich and complex.

4 Case Studies

We performed some case studies whose main objective was to evaluate the proposed architecture suitability to provenance collection and management in the context of scientific experiments processed in a collaborative research environment. Moreover, these case studies were used to evaluate scientific workflows performance after instrumentation by *SciProv* to capture provenance according to OPM model. To achieve proposed objectives, we used available resources at DNA Data Bank of Japan (DDBJ) [23], member of International Nucleotide Sequence Database Collaboration (INSDC). Specifically, these case studies used the interface called Web API for Biology (WABI)[1] which gathers DDBJ web services specifications, tutorials, research tools, templates, and workflows built from the available infrastructure.

In developing the case studies, we choosed *getEntry* web service, which performs a comprehensive search in DDBJ and returns a broad range of information relating to genetic sequencing of a specific organism according to formulated search criteria. The first case study uses Kepler SWfMS. Figure 5 shows provenance graph from the scientific workflow modeled.

Besides demonstrating the applicability of *SciProv* in a scenario consistent with the proposal, one objective of case studies is to formulate a semantic search to extract information that is not explicitly represented and therefore must be inferred from the OPM ontology specialized for scientific workflows domain.

One of the refinements proposed in OPM specialization is related to object properties named *triggeredProcess* and *triggeredByProcess*. These properties represent relationships between classes *Triggered* and *Process* and, in accordance with the original OPM ontology, are not considered inverse and/or transitive. From a semantic search in SPARQL, the *SciProv* is able to return information inferred from knowledge about OPM specialized model and which are not explicitly represented in provenance graph as highlighted in Figure 6.

The second case study uses Taverna to model a scientific workflow with identical functionality to the previous one. The objective was to verify that *SciProv* is able to generate equivalent provenance metadata from different SWfMS. Results obtained in both case studies were identical with regard to provenance graph and semantic query.

The third case study aimed at presenting a situation that may be interesting in collaborative environments. Specifically, what one sees is the design of workflows modeled together using two or more SWfMS. Thus, one can consider that part of the work is developed in a specific research center using Taverna and another part is modeled in a separate location using Kepler. In Figure 7, provenance metadata shown in the graph until the artifact called "DDBJ Data" were obtained from Taverna. All other metadata represented in a light gray level, were obtained from Kepler. This approach facilitates the understanding of metadata origin in a jointly modeled scientific workflow. We can observe that the graph obtained remains consistent with previously presented in Figure 6.

---

[1] `http://xml.nig.ac.jp/index.html`

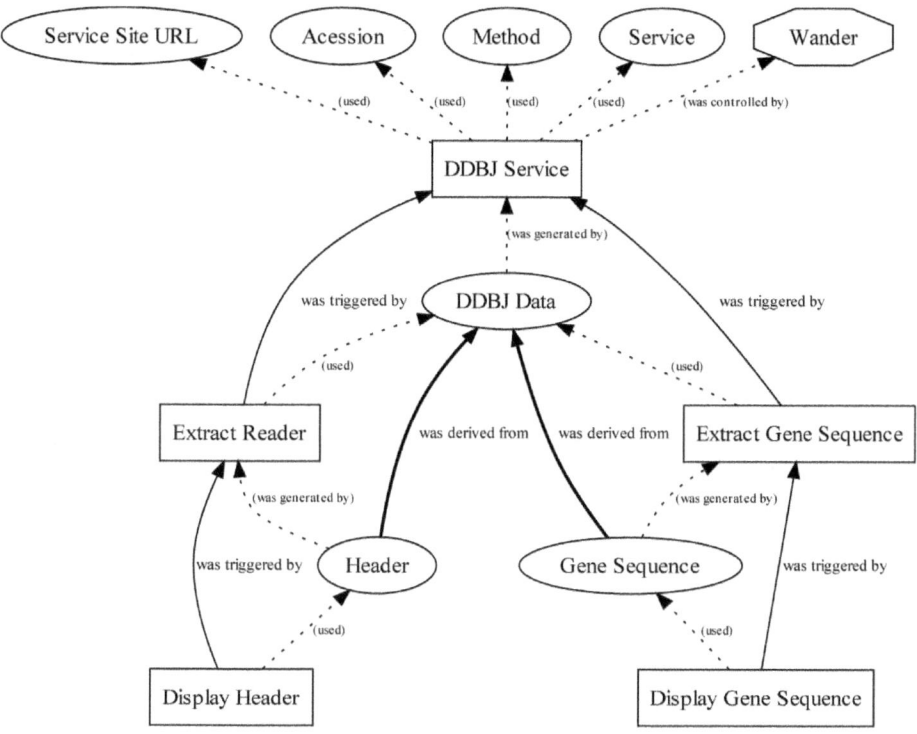

Fig. 5. Scientific workflow provenance graph (Kepler version)

fromProcess	fromValue	toProcess	toValue
id:p1	DDBJ Service	id:p5	Display Gene Sequence
id:p1	DDBJ Service	id:p4	Display Header
id:p1	DDBJ Service	id:p3	Extract Gene Sequence
id:p1	DDBJ Service	id:p2	Extract Reader
id:p2	Extract Reader	id:p4	Display Header
id:p3	Extract Gene Sequence	id:p5	Display Gene Sequence

Fig. 6. Semantics query results in OPM provenance graph

The following case study aimed to evaluate the applicability of resources provided by OPM annotation framework for semantic queries composition based on this kind of metadata. A new annotation type was included in its framework. For each *Process* entity, the annotation should record the importance of their correct execution for the workflow in order to be considered fully processed (Figure 8). This information is relevant considering scientific experiments performed on

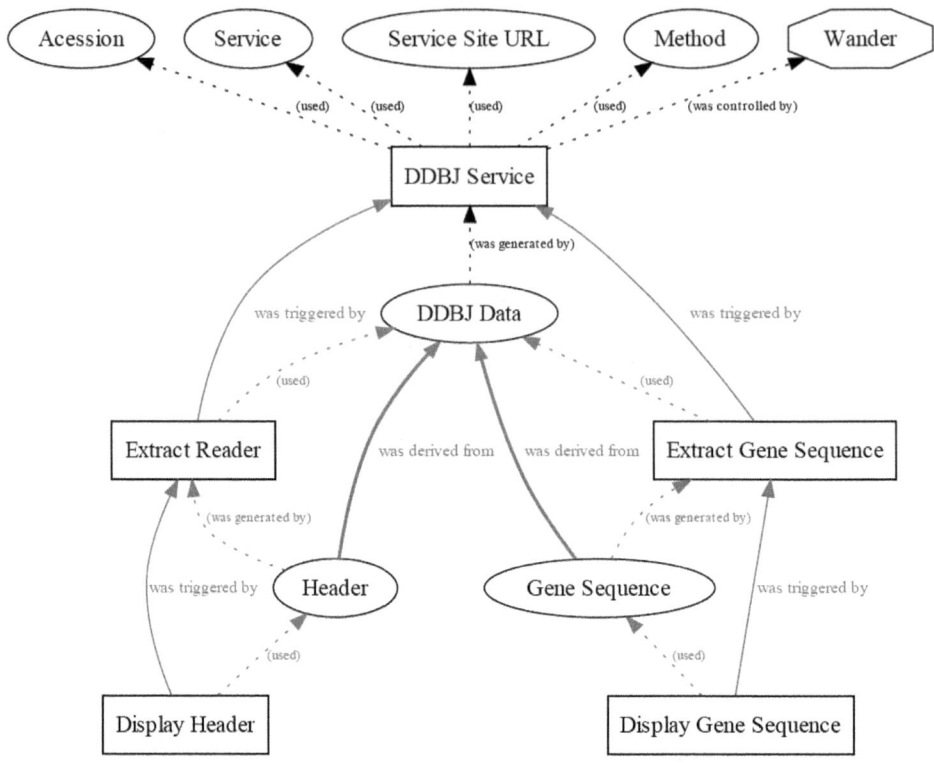

Fig. 7. Jointly modeled provenance graph: Kepler (gray) and Taverna (black)

a computational grid. For example, a web service may become unavailable or process may return with error. Thus, it is interesting to define which processes need to be successfully completed so that the workflow can present the expected results.

This case study included a semantic search in SPARQL to return *Process* entities that have been annotated with the value *High*. The result was consistent with the provenance graph shown in Figure 8.

The final case study examined a further semantic enrichment possibility. The *SciProv* allows the composition of a *TBox* drawing on both OPM model ontology — plus scientific workflows specialization — and a second ontology for the science domain addressed (Figure 9).

In this case study we used an ontology to represent a partial taxonomy of Kingdom *Bacteria*. This ontology has been incorporated into *SciProv* TBox and a semantic search was made to the provenance graph. Figure 10 shows consistently *Synechococcus elongatus* Family and Genus in accordance with Cavalier-Smith classification [4].

However, it is considered that the process of adding two or more different ontologies to *SciProv* TBox may require prior processing of ontologies alignment.

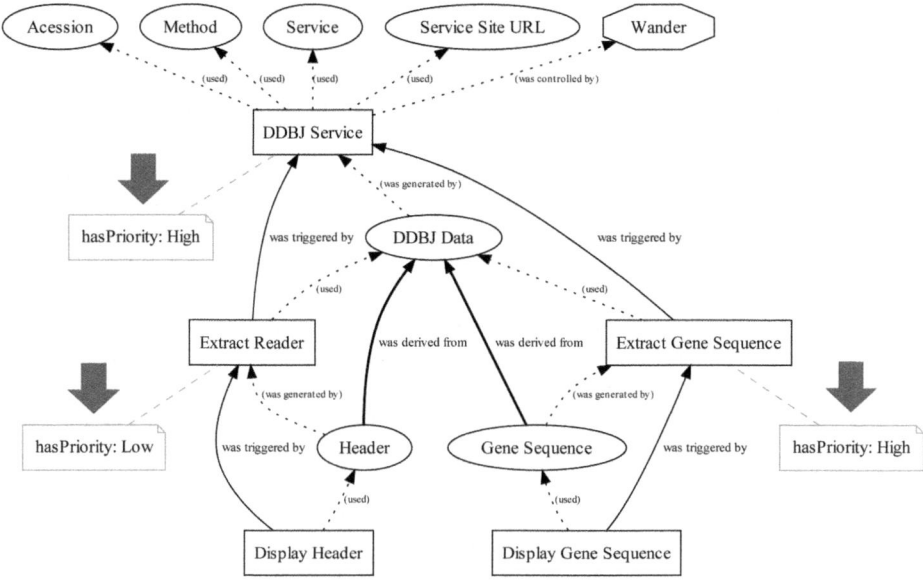

Fig. 8. Provenance graph with annotations

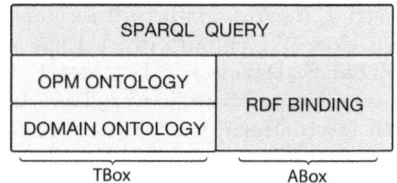

Fig. 9. TBox with both OPM and domain ontologies

artifact	acession	organism	genus	family
id:a3	AB000100	http://www.semanticweb.org/ontologies/2010/11/opm2.owl#Synechococcus_elongatus	http://www.semanticweb.org/ontologies/2010/11/opm2.owl#Synechococcus	http://www.semanticweb.org/ontologies/2010/11/opm2.owl#Synechococcaceare

Fig. 10. Semantic query results for *Synechococcus elongatus*

5   Conclusion and Future Work

Based on case studies presented, it can be considered that *SciProv* is adequate
for data provenance considering the presented context. We can emphasize that
one of *SciProv* key features is the possibility of extracting information inferred
from semantic queries and the ability to collect and manage provenance metadata independently of any specific formats. Research on data provenance is an

open issue and, in the context of e-Science, some questions may be raised as suggestions for future work related to *SciProv*.

- Refinement of the proposed specialization for OPM ontology, in order to represent more consistently and comprehensively the provenance graph in scientific workflows domain.
- Refinement of instrumentation mechanism, in order to optimize performance of metadata provenance collection and persistence. A simpler alternative would be sending the metadata in batch to the repository. Another option would be lineage distributed storage on different nodes of the computational grid.
- Evaluation of alternative technologies for provenance semantic query. It can be argued that SPARQL language, although considered a standard by W3C, imposes limitations on the use of *SciProv* by scientific community due to the complexity. One possibility being considered is to use a graphical representation of lineage graph to formulate queries from a visual and intuitive interface.
- Maintain continued adherence to OPM abstract model, which is rapidly developing and updating. OPM has tried to respond to various data provenance challenges. It is important that *SciProv* be able to incorporate model features that came to be included or enhanced.

References

1. Barga, R.S., Digiampietri, L.A.: Automatic capture and efficient storage of e-science experiment provenance. Concur. Comput.: Pract. Exper. 20(5), 419–429 (2008)
2. Biton, O., Cohen-Boulakia, S., Davidson, S.B., Hara, C.S.: Querying and managing provenance through user views in scientific workflows. In: ICDE 2008: Proceedings of the 2008 IEEE 24th International Conference on Data Engineering, pp. 1072–1081. IEEE Computer Society, Washington, DC, USA (2008)
3. Buneman, P., Khanna, S., Tan, W.C.: Why and where: A characterization of data provenance. In: Van den Bussche, J., Vianu, V. (eds.) ICDT 2001. LNCS, vol. 1973, pp. 316–330. Springer, Heidelberg (2000)
4. Cavalier-Smith, T.: Only six kingdoms of life. Proceedings of the Royal Society B Biological Sciences 271(1545), 1251–1262 (2004)
5. Cruz, S.M.S.d., Campos, M.L.M., Mattoso, M.: Towards a taxonomy of provenance in scientific workflow management systems. In: SERVICES 2009: Proceedings of the 2009 Congress on Services - I. IEEE Computer Society, Washington, DC, USA (2009)
6. Davidson, S.B., Freire, J.: Provenance and scientific workflows: challenges and opportunities. In: SIGMOD 2008: Proceedings of the 2008 ACM SIGMOD International Conference on Management of Data, pp. 1345–1350. ACM, New York (2008)
7. Digiampietri, L., Medeiros, C., Setúbal, J.: A framework based in web services orchestration bioinformatics workflow management. Genetics and Molecular Research 4(3), 535–542 (2005)
8. Factor, M., Henis, E., Naor, D., Rabinovici-Cohen, S., Reshef, P., Ronen, S., Michetti, G., Guercio, M.: Authenticity and provenance in long term digital preservation: modeling and implementation in preservation aware storage. In: First Workshop on Theory and Practice of Provenance, pp. 6:1–6:10. USENIX Association, Berkeley (2009)

9. Freire, J., Koop, D., Santos, E., Silva, C.T.: Provenance for computational tasks: A survey. Computing in Science and Enginnering 10(3), 11–21 (2008)
10. Freire, J., Silva, C.T.: Towards enabling social analysis of scientific data. In: CHI Social Data Analysis Workshop, Florence, Italy (2008)
11. Freire, J., Silva, C.T., Callahan, S.P., Santos, E., Scheidegger, C.E., Vo, H.T.: Managing rapidly-evolving scientific workflows. In: Moreau, L., Foster, I. (eds.) IPAW 2006. LNCS, vol. 4145, pp. 10–18. Springer, Heidelberg (2006)
12. Golbeck, J., Hendler, J.: A semantic web approach to the provenance challenge. Concurr. Comput.: Pract. Exper. 20(5), 431–439 (2008)
13. Groth, P., Deelman, E., Juve, G., Mehta, G., Berriman, B.: Pipeline-centric provenance model. In: WORKS 2009: Proceedings of the 4th Workshop on Workflows in Support of Large-Scale Science, pp. 1–8. ACM, New York (2009)
14. Groth, P., Jiang, S., Miles, S., Munroe, S., Tan, V., Tsasakou, S., Moreau, L.: An architecture for provenance systems. Tech. Rep. D3.1.1 Final Architecture v.0.6, EU Provenance Project, Southampton, UK (2006)
15. Ludäscher, B., Altintas, I., Berkley, C., Higgins, D., Jaeger, E., Jones, M., Lee, E.A., Tao, J., Zhao, Y.: Scientific workflow management and the kepler system: Research articles. Concurr. Comput.: Pract. Exper. 18(10), 1039–1065 (2006)
16. Marinho, A., Murta, L., Werner, C., Braganholo, V., da Cruz, S.M.S., Ogasawara, E., Mattoso, M.: Managing provenance in scientific workflows with provmanager. In: 22nd International Symposium on Computer Architecture and High Performance Computing. LNCC (2010)
17. Moreau, L., Clifford, B., Freire, J., Gil, Y., Groth, P., Futrelle, J., Kwasnikowska, N., Miles, S., Missier, P., Myers, J., Simmhan, Y., Stephan, E., Bussche, J.: The open provenance model core specification (v1.1). Future Generation Computer System (2010) (in press, corrected proof)
18. Moreau, L., Freire, J., Futrelle, J., Mcgrath, R.E., Myers, J., Paulson, P.: The open provenance model: An overview, pp. 323–326 (2008)
19. Munroe, S., Miles, S., Moreau, L., Vázquez-Salceda, J.: Prime: a software engineering methodology for developing provenance-aware applications. In: SEM 2006: Proceedings of the 6th International Workshop on Software Engineering and Middleware, pp. 39–46. ACM, New York (2006)
20. Oinn, T., Li, P., Kell, D.B., Goble, C., Goderis, A., Greenwood, M., Hull, D., Stevens, R., Turi, D., Zhao, J.: Taverna/mygrid: Aligning a workflow system with the life sciences community. In: Workflows for e-Science: Scientific Workflows for Grids, ch. Part III, pp. 300–319. Springer, London (2007)
21. Olson, G.M.: The next generation of science collaboratories. In: CTS 2009: Proceedings of the 2009 International Symposium on Collaborative Technologies and Systems, pp. xv–xvi.. IEEE Computer Society, Washington, DC, USA (2009)
22. Simmhan, Y.L., Plale, B., Gannon, D.: A framework for collecting provenance in data-centric scientific workflows. In: ICWS 2006: Proceedings of the IEEE International Conference on Web Services, pp. 427–436. IEEE Computer Society, Washington, DC, USA (2006)
23. Tateno, Y., Imanishi, T., Miyazaki, S., Fukami-Kobayashi, K., Saitou, N., Sugawara, H., Gojobori, T.: Dna data bank of japan (ddbj) for genome scale research in life science. Nucleic Acids Research 30(1), 27–30 (2002)
24. Zhao, J., Goble, C., Stevens, R., Turi, D.: Mining taverna's semantic web of provenance. Concurr. Comput.: Pract. Exper. 20(5), 463–472 (2008)

Integration of Procedural Knowledge in Multi-Agent Systems in Medicine

Lenka Lhotská, Branislav Bosansky, and Jaromir Dolezal

Dept. of Cybernetics, Czech Technical University in Prague,
Technicka 2, 166 27 Prague 6, Czech Republic

Abstract. In this paper we present two developed multi-agent architectures. The first one models knowledge-based support of the eServices in the domain of home care, while the second one represents a general model that uses procedural knowledge in the form of organizational processes and formalized medical guidelines in a decision support system. We show that integration of both architectures is feasible and moreover, it can enhance functionality of the home care system.

Keywords: multi-agent system, home care, formalized medical guidelines, process modelling, ontology, organizational process.

1 Introduction

Medical environments use to involve high complex structures of interacting processes and professionals where a high quantity of information is managed and exchanged. In addition, all healthcare environments can be considered distributed. The services usually involve professionals from different institutions (hospital, social work organizations, etc.) structurally independent, which must interact around any particular patient, and which usually are located in different physical places having their own and independent information systems. Thus the applications must work in a highly distributed environment using several different types of communication and end-user devices. Due to the distributive nature of decision support systems in health care, multi-agent architecture is often employed. Moreover, the multi-agent architecture is particularly suitable for the domain of home care. Therefore, while building the decision support system for home care, K4Care, we followed the multi-agent paradigm.

Each decision support system has to work with different types of knowledge. On one hand, the system must consider and analyse the necessary medical knowledge, however, on the other hand, each system must take into consideration the actual institution, in which it is deployed, and follow the local specifics. Both of these types of knowledge are often captured as more general procedures – or more precisely, they are formalized as a procedural knowledge. Medical knowledge is often formalized in the form of medical guidelines that contains recommended procedures for diagnosing specific diseases and/or their treatment. Formalization of the organizational processes on the other hand follows from the domain of business processes and uses general methods of business process modelling.

C. Böhm et al. (Eds.): ITBAM 2011, LNCS 6865, pp. 82–95, 2011.

In this paper we describe how the multi-agent architecture of the K4Care system can be combined with a general multi-agent architecture that works with procedural knowledge. Firstly, we give more detailed description of the K4Care system that offers support of eServices in the home care, following by a description of a general multi-agent architecture that can use formalized procedural knowledge. Finally, we show how these two architectures can be combined. In this work we continue our preliminary analysis of possibilities of such integration given in [1] by providing more detailed analysis of the K4Care platform and focus on the concept of services in the home care system and their impact on the integration.

2 Knowledge-Based Home Care eServices

The care of chronically ill patients involves lifelong treatment under continuous expert supervision. Admission to hospital and residential facilities can be unnecessary and even counterproductive, and could saturate national health services and increase health care costs.

K4Care Platform [2] is a software system providing Knowledge-Based HomeCare eServices. It is a response to the needs of the increasing number of senior population requiring a personalized HomeCare (HC) assistance. The system integrates knowledge and skills of specialized centres and professionals of several EU countries. The knowledge and skills are incorporated in an intelligent web platform in order to provide e-services to health professionals, patients, and citizens in general. It offers a unique solution that integrates features of a healthcare information system with a decision support system and works in highly distributed environment.

The K4Care platform serves a broad range of HC providers such as local health units integrated with social services of municipalities, specialized HC centres, other organizations of care or social support as well as specialized care (e.g. rehabilitation, oncology, etc.).

K4Care provides medical doctors with following functions:

- Creating record of a new patient;
- Storage and retrieval of patient records using defined selection criteria (e.g. recently examined patients, greatest interval from the last visit, the same diagnosis, etc.);
- Inspection of formal intervention plans (developed from medical guidelines), the right to develop new formal intervention plans have only some qualified users;
- Development of individual intervention plan (IIP) for each patient based on his/her diagnosis (or combination of diagnoses) and previous treatment. These plans are successively stored in the patient record.

Other care givers are provided with relevant data and information about the patients and are allowed to insert data specified in medical doctors´ requests.

2.1 Home Care Model

The K4Care Platform is developed around the HomeCare model [3] that defines all entities, their relations and interactions in the course of home care activities. The entities (doctors, nurses, other care providers, patients, etc.) are represented by actors.

Their behaviour is formalized in so-called Actor Profile Ontologies. In this way actions and relations of each actor (user) in the home care framework are described.

The model proposes that the services are distributed by local health units and integrated with the social services of municipalities, and possibly other organizations of care or social support. It is aimed at providing the patient with the necessary sanitary and social support to be treated at home; in these terms, it is easily adaptable to those socio-sanitary systems which provide the so called "unique access" for both social and sanitary services, unifying and simplifying procedures of admission to the services. The model could represent an incentive and facilitate the shift towards such an approach. To accomplish this duty, the Model is designed give priority to the support of the HCP, his/her relatives and Family Doctors (FD) as well. The described model was developed in frame of the K4Care project and thus we will further use the notation K4Care Model.

As shown in Figure 1, the K4Care Model is based on a nuclear structure (HCNS) which comprises the minimum number of common elements needed to provide a basic HC service. The HCNS can be extended with an optional number of accessory services (HCAS) that can be modularly added to the nuclear structure. These services respond to specialized cares, specific needs, opportunities, means, etc. of either the users of the K4Care model or the health-care community where the model is applied.

The distinction between the HCNS and the complementary HCAS's must be interpreted as a way of introducing flexibility and adaptability in the K4Care model and also as an attempt to provide practical suggestions for standards to be used when projecting and realizing new services in largely different contexts.

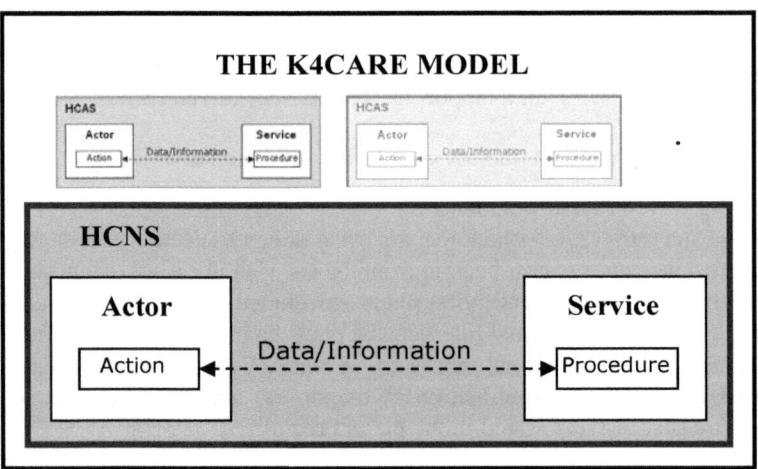

Fig. 1. The K4Care Model Architecture for HC

2.2 Home Care Structures

Each of the HC structures (i.e. HCNS and HCAS's) consists of the same components: actors, actions, services, procedures, and information interface (data and devices).

- *Actors* are all the sort of human figures included in the structure of HC.
- *Professional Actions and Liabilities* are the actions each actor performs to provide a service within the HC structure.
- *Services* are all the utilities provided by the HC structure for the care of the HCP.
- *Procedures* are the chains of events that lead an actor in performing actions to provide services.
- *Information* is the documents required and produced by the actors to provide services in the HC structure.

The Home Care Nuclear Structure of the K4Care HC Model comprises the minimum elements needed to provide a basic HC service. These elements are the actors involved in HC, the actions and liabilities of such actors, the services available in the HC model, the procedures, and the information to provide such services.

In HC there are several people interacting: patients, relatives, physicians, social assistants, nurses, rehabilitation professionals, informal care givers, citizens, social organisms, etc. In the HCNS, these individuals are the members of three different groups of HC actors that this section describes. These groups are the patient; the stable members of HCNS (the family doctor, the physician in charge of HC, the head nurse, the nurse, the social worker, each of them present in the HCNS); the additional care givers. The family doctor, the physician in charge of HC, the head nurse, and the social worker join in a temporary structure – the Evaluation Unit – devoted to assess the patient's problems and needs.

The HCNS provides a set of services for the care of HCP. These services are classified into Access, Patient Care, and Information services. Access services see the actors of the HCNS as elements of the K4Care model and they address issues like patient's admission and discharge from the HC model. Patient Care embraces the most complex services of the HC model by considering all the levels of care of the patient as part of the HCNS. Finally, Information services cover the needs of information that the HCNS actors require in the K4Care model.

2.3 Implementation

The basic ICT components of the system include a data repository, the rules engine, several knowledge sources, some data mining tools and a human–computer interface, allowing remote access.

Since K4Care handles sensitive medical data, all the important security issues are properly addressed, namely authorization and authentication of the users, access rights, secure transmission of data, private data protection.

The platform offers an implementation of the general HC model through a set of knowledge structures and completely independent execution logics through several cutting edge technologies such as JADE Intelligent Agents, Ajax-based web interface, OWL ontologies and XML-based Electronic Health Care Record. The implementation is OS-independent and easy to manage. The components are highly reusable and the system's behavior can be easily modified, adapted or personalized to the characteristics of a particular healthcare organization.

The system is composed of several layers, namely graphical user interface and platform itself, data abstraction layer, and knowledge layer (see Figure 2).

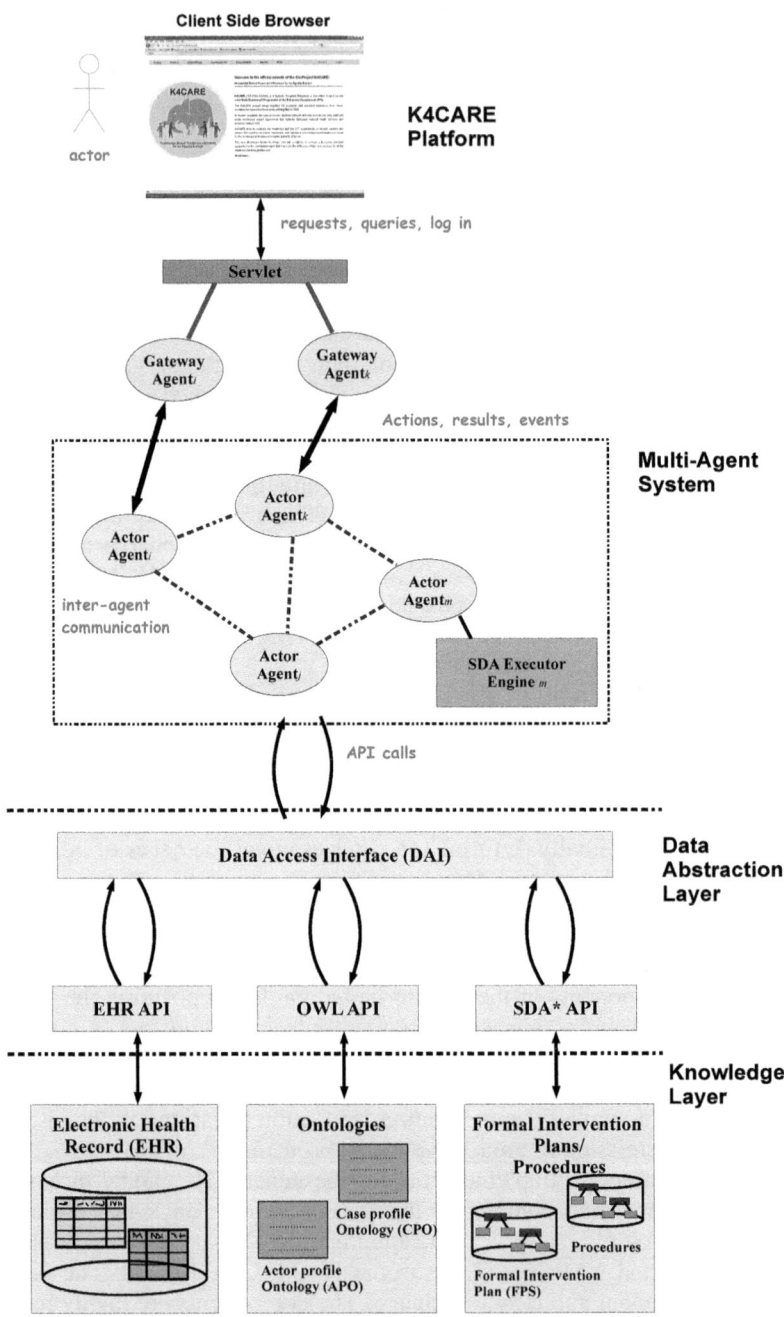

Fig. 2. Architecture of the K4Care system

The human–computer interface is a web-based application accessible at the client side by a Web browser either on a PC or laptop, or a PDA. The web interface is made simple, easy and intuitively to use and it is constructed based on each user role. The user is represented within the system by a permanent agent (Actor Agent) that incorporates all details about his/her roles, permissions and pending actions. Actor Agents are responsible of coordinating their activities for the successful and efficient execution of administrative procedures and personalized medical treatments (IIPs).

The actors communicate with the platform through the web interface which permits authorized users to request services, see pending actions, look up the electronic health care record of the patients, fill in documents, update their profile, work as part of a group, etc. All these interaction are implemented as various types of documents in a standard form. This is made for the actors to concentrate on their part of the care, while still seeing it in a wider context of care.

The Knowledge Layer includes all the data and knowledge sources required by the platform. It contains an Electronic Health Care Record that stores patient records with personal information, medical visits and ongoing treatments. It is implemented by a collection of XML standard documents which structure medical data properly [4].

The Data Abstraction Layer (DAL) provides Java-based methods that allow the K4Care platform entities to retrieve the data and knowledge that they need to perform their tasks.

This layer offers a wide set of high-level queries that provide transparency between the data and the knowledge, their representation and their use in the platform.

The next layer is the platform itself comprising of agents, servlets, and web interface.

All the data is stored in modern portable formats such as properly structured XML format of the documents, HL-7 compatible database EHCR record, OWL (Web ontology language) for the knowledge, permissions and schematic model of the clinical decision pathways in a SDA (States, Decisions and Actions) format [5] to guarantee interoperability with other systems. International codification systems are used for classification of diseases, etc.

It is important that the knowledge is not hard coded and thus it can be easily adopted for concrete needs of specialized care. The general approach also enables the use of the system in other areas of medical care and other settings of facilities since the K4Care platform is not just a implemented knowledge that was gathered during the development but rather a open system providing a formalization, repository and tools to import, input, execute, evaluate and extend the knowledge during utilization of the system.

3 Using Business Processes in Health Care

The specification of work practice using business processes is for a long time a fundamental topic for most companies regardless to their field of industry. They define duties of employees, work procedures in most common situations, and possible cooperation throughout a team, departments, or even several companies. During this time, a lot of applications based on processes have been developed either to help create and model the business processes as a part of business process management, or.

We understand the term process to be a sequence of activities, states, decision points, and steps splitting the sequence or synchronization points. Processes in health care can be divided into two basic categories – the organizational processes and the medical treatment processes.

In the analysis of the first category, we are interested in applications of processes in information systems in order for their automatic execution, monitoring, or guidance of users. This task is generally supported by several existing tools (e.g. based on BPEL[6], or EPC[7] languages etc.). However, there are several complications while business processes in the domain of healthcare and there are several studies (e.g. in [8], [9], [10]) that practically examine problems of applying process modelling or usage of workflow management systems within the health care domain. There is a mutual agreement that a successful implementation of this approach can improve the quality of health care, reduce the time of necessary hospitalization and lead to the reduction of costs. Main reason lies in variety of legal restrictions resulting from the basic principle that patient health has the highest priority over the healthcare processes. Secondly, processes in the health care domain severely depend on a concrete patient, his/her current health conditions and further treatment, which is based on complex medical knowledge. Therefore multiple exceptions would have to be considered while modelling or using created processes. In spite of these obstacles in implementation of the appropriate support by formalized processes there is a consensus that such programs would improve the quality of the health care. Mainly, the potential is seen in monitoring and critiquing systems that would transparently control performed actions and alert in case of possible danger – when a necessary examination before starting a treatment was omitted, or some data are missing in the patient electronic health record (EHR) in case of his/her transport to another facility.

Medical treatment processes are captured as medical guidelines, which are for some time now a part of process of standardization of medical treatment. Guidelines contain recommended actions, directions, and principles for treatment of specific diseases approved by appropriate expert committees. They act as pillars of evidence-based medicine helping that way physicians with complicated clinical decisions. There are several workgroups and several languages (PROforma[11], Asbru[12], GLIF[13], etc.) that capture the knowledge of a textual medical guideline into an electronic and structured form. All of them focus on specific parts – e.g. logical formalism in Asbru, or automatic execution and retrieval of patient's data in GLIF. And finally, they are all based on a process-oriented approach.

3.1 Process-Based Multi-Agent System

In this section we briefly describe the architecture and the functioning of the multi-agent system that could be used as a basis for the critiquing decision support system. The scheme of the developed architecture (termed ProMA [1,15]) is depicted in Figure 3 showing the basic types of agents. The architecture is more general and it can be used not only for simulating the processes, but also for their monitoring. Hence we describe these agents and their purpose more in detail in the role of process simulation.

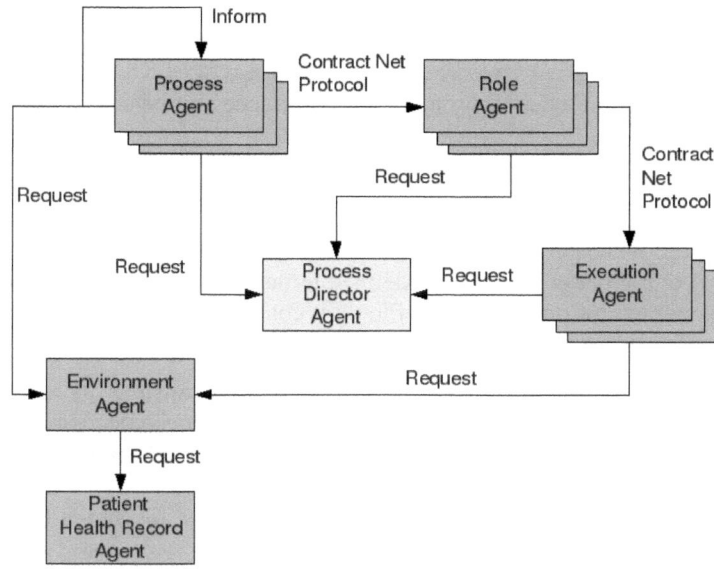

Fig. 3. ProMA - Architecture of a multi-agent system able to handle processes

3.2 Environment Agent

Environment Agent represents the environment of the simulation. With respect to the level of detail that we want to model using the system, the environment could represent a virtual world (e.g. database of patients' health records, a whole department of a hospital, etc.) with existing objects (e.g. laboratory results, RTG or EEG machines, beds, etc.). As it is shown in the figure, agent managing the patients' EHR is requested by the Environment Agent as patients' data certainly are a part of environment. However, due to simulation purposes there can be necessary to distinguish between real data stored in the record (Patient Health Record Agent) and data used for the simulation (Environment Agent).

3.3 Execution Agent

Execution Agents (EAs) represent concrete physicians, nurses, patients, or possibly other employees of the facility that are involved in the processes. Each EA has several pre-defined rules, that for basic behaviour in the environment. Then, for each activity that the agent (hence the represented person) can participate in, one additional rule is generated. These rules can be activated or deactivated by a Role Agent (see below), while each Execution Agent autonomously chooses which of the activated processes it will execute based on the priority in which the rules are ordered.

3.4 Role Agents

Role Agents (RAs) represent the roles in the environment (i.e. general roles for patient, nurse, physician, etc.). RA receives the proposal from a Process Agent (see below) and is responsible to select appropriate Execution Agent(s) (EA) with associated (or more specific) role, maintain its progress and find another EA in case the current one needs to suspend its participation on currently executed activity (e.g. one physician is needed in a case with a higher importance and he/she can be replaced in the current one, etc.).

3.5 Process Agent

For every step in the process notation (i.e. for each activity, event, decision point, etc.) there is one Process Agent (PA) defined in the architecture. The PA is responsible for a proper execution of the step. Firstly it controls whether the initial conditions for the process are met:

- whether the predecessing PA(s) has successfully finished its execution (i.e. whether the PA has received the inform message from its predecessor(s));
- whether all input objects have the needed values (the PA uses a simple request protocol to retrieve these data from Environment Agent);
- whether there exist appropriate agents that will execute this action (the PA uses a CNP for those RAs that are connected with this activity).

When all mandatory conditions hold, the PA starts the execution of the step (e.g. simulation of an activity, calculation, decision process, etc.) and after successful finish, the PA is responsible for notifying the Environment Agent about the results of the activity (using simple request protocol) and the next succeeding Process Agent about the successful finish (using simple inform protocol).

3.6 Process Director Agent

Last agent is an auxiliary agent that reads the formalized processes and answers other agents to their requests regarding information about the processes (e.g. input data, predecessors, etc.)

4 Integration of Process-Based Architecture into the K4Care System

In this section we describe the integration of the ProMA architecture with the architecture of K4Care system. Firstly we describe conceptual and technical changes that need to be performed, following by proposition of further capabilities that the enhanced K4Care model can provide. The reason is in full utilization and implementation of the concept of eServices in the K4care model. Primal improvements are noticeable while creating a new IIP for a patient by a human expert that receives much better support in his/her decision process. However, further possible improvements can be developed as we discuss in the next section.

4.1 Environment

We need to take the K4Care architecture into consideration (see Figure 2). First of all, there is a need for unification of existing functionality of the K4Care system with the

ProMA architecture. The connection to the health record data and procedural knowledge of the multi-agent system represented in ProMA as the Patient Health Record Agent and the Process Director Agent is in the K4Care realized by the Data Abstraction Layer (DAL). This layer provides Application Program Interface (API) for desired request such as about patient's data from the health record or guidelines formalized in SDA.

However, the Environment Agent must still be present in order to temporarily store simulation-based data (that are not persistent). Due to the fact that the DAL is in the current version not able to satisfy partial request on the structure of the formalized processes as performed by Process or Role Agents in ProMA (DAL was designed only for retrieving complete documents), hence both the Process Director Agent and Patient Health Record Agent should be created in the multi-agent system as well in order to respond to these requests.

4.2 Execution Agents

The Execution Agents (EAs) in the ProMA conceptually correspond to Actor Agents in the K4Care system – they represent the actual users of the system and executors of the actions, receive requests to perform specific actions and after real actors finish, Actor Agents store the output data into related documents. This corresponds with the EAs more general reactive behavioural model. However, in order to fit the ProMA model completely, Actor Agents need to be able to communicate with Role Agents (RAs), and autonomously react on their requests for participation in an activity. Note, that in the current version of the K4Care system providing this response is solely based on human user activity. However, due to the persistent presence of each Actor Agent in the system, their functionality can easily be enhanced with the ability to response to the contract-net protocol (CNP) initiated by Role Agents (RAs). The response can be based on currently associated tasks and/or availability of the related user (e.g. if a medical specialist has left the facility for a longer time, the Actor Agent could posses this type of information and automatically deny his/her participation on proposed activities). Moreover, the SDA concepts typically have the notion of time-related information stored (i.e. time interval and step of the occurrence for a state or an activity), hence Actor Agents can provide much more sophisticated answers based on currently planned actions, and the associated IIPs, and act as a basic time-scheduling algorithm easing the human expert when creating a new IIP for a patient (see the next section for new possible features).

4.3 Role Agents

The rest of the types of agents presented in the ProMA architecture do not have their alternative in the K4Care model, hence they have to be created. In the following we describe their full integration into K4Care system. Note that for Role Agents (RAs), neither Process Agents (PAs) is necessary to be persistent in the system within proposed use-case of creating an IIP (only relevant roles and processes should be created based on requested FIPs).

Role Agents (RAs) conceptually correspond to types of actors in the K4Care system, however, there are no specific agents present yet. Hence, for each actor type we create one RA as follows: Patient, Family Doctor, Head Nurse, Physician in

Charge, Social Worker, Nurse, Specialist Physician, Social Operator, Continuous Care Provider, and Informal Care Giver. Moreover, these types are already in the K4Care system organized into groups: Evaluation Unit (where Family Doctor, Head Nurse, Physician in Charge and Social Worker belong), and Additional Care Givers (where Specialist Physician, Social Operator, Continuous Care Provider, and Informal Care Giver belong). Therefore, for each of these groups one RA is created as well and a hierarchy between them can be established. Moreover, further hierarchy can be seen in various activities (most of the activities of a nurse can be executed by a head nurse as well, etc.)

The concept of Execution (EAs) and Role Agents (RAs) particularly complies with the concept of services and nuclear structure used in the K4Care system. The providers of home care services are associated with the procedures they offer to the health care specialist. Similarly, the RAs and EAs are connected with corresponding processes in the SDA model and they can offer their services to the Process Agents, which can then choose a provider that best fit the current situation.

4.4   Process Agents

Finally, for each state, decision or activity in requested FIP SDA models one Process Agent (PA) is created in the multi-agent system. When the IIP is being created by the user, for each step one PA is created as well. The collective behaviour of the PAs during the execution of the IIP by health care specialists correspond to the behaviour implemented the SDA Execution Engine. However, the advantage of using the specific PA for each step in the SDA procedure is the creating the IIP by merging the requested FIPs, which can be partially automated, hence offer some preliminary result that a human expert can use and further improve according to his/her knowledge.

All of the characteristics of the SDA steps in the K4Care system are formalized using exhaustive ontology structure, hence we can utilize this information and try to automatically merge some of the steps from requested FIPs in order simplify the final model. The equivalence of several steps can be identified by appropriate Process Agents comparing associated terms and context. If they match, a single Process Agent can be created in the IIP with the union of needed documents and output data (e.g. a blood-analysis can be performed only once with a union of desired information). Although the idea of semi-automated merging whole SDA procedures was briefly introduced in [13], we propose merging of separate steps of the procedures using autonomous agents. This can be further improved by using a basic simulation (will be described below) of all of the processes concurrently (which ProMA offers). The PAs can identify the order in which they can be executed based on fulfilment their prerequisites. By the term "basic simulation" we mean a simulation, where the PAs try to simulate associated steps by temporarily filling necessary data fields of the patients' data profile with arbitrary values and analyse which of following PAs have fulfilled their prerequisites. We further improve this insight in the next section by more advanced simulation.

Finally, using the CNP Protocol, Process Agents (PAs) can delegate a problem of selecting appropriate health care specialist to associated Role Agents (RAs). There is a key feature of the RAs in the K4Care system – requesting a set of Actor Agents must obey security restrictions based on the current step in the IIP and current patient.

Based on returned responses the RAs can offer to the PAs a best fit, or a sorted list of possible candidates for a human expert selection.

5 Further Improvements and Discussion

In the previous section we described integration of the ProMA architecture into the K4Care system, which required only minimal conceptual and model changes. The key advantage of this integration is in opening new possibilities for the K4Care system easily realized using ProMA. These, however, would require additional knowledge about the activities supported by the system functions.

The key factor is to use advanced simulation during the progress of developing a new IIP. The advantage of the K4Care system is that both type of processes – as organizational, as well as formalized medical guidelines are stored using the same formalism SDA and are viewed and considered together while tailoring a specific IIP for a patient. As showed in the previous section, even basic simulation can find some constraints that can be visualized and ease the process of creating the IIP for the human expert.

Let us assume additional information available in the system. First of all, we can enhance ontology for each action with additional properties, such as duration and type of progress function (see [15] for more details). Secondly, map of a local area that the system is used in can be integrated into the system as a virtual environment. Finally, we can request to access other relevant patients data from the EHCR database during the process of creating the IIP (while keeping the security restrictions). All these additional parameters allow us to simulate modelled procedures on more detailed level. Based on a proposition of the IIP, its simulation can be run resulting in an evaluation of the current version of the IIP and associated actors.

The simulation advance as in ProMA, which means each of Process Agents is trying to successfully simulate associated step by selecting appropriate Role and Actor Agents. As the Actor Agents are limited by other tasks already set in the system (i.e. there are several IIPs that are simulating at one time), they have to select the actions in which they currently participate according to their capabilities and priority. Let us demonstrate it on an example: we need to associate a nurse with an activity N.15 Blood sugar measurement at patient's home. We have a nurse Alice and a head nurse Eve. Nurse Alice has already planned a single task on the other side of the town, head nurse Eve has planned two shorter tasks in the centre. Note, that we use actors with different roles, however, as there is hierarchy allowed within these roles and activity (i.e. a nurse is a more general role than the head nurse role). The ProMA architecture is able to handle this hierarchy and the RA corresponding to the nurse role would also send the proposition to RA corresponding to the head nurse role. In order to select an appropriate actor for this activity we use a multi-agent simulation in the virtual environment. That means the Actor Agents simulate execution of their already planned tasks within this environment. When the request for the N.15 activity is proposed by the RA associated with the nurse role, both Actor Agents are in different locations in the virtual environment and based on their other tasks they can estimate the time of completion of the proposed task. Hence, in this simplified case, Eve would be offered for a human expert as an appropriate actor for this activity. However, if the proposed activity would be with higher priority for Alice, and her current task could be temporarily suspended, she could be selected. In order the simulation of the steps be as accurate as possible, PAs must have access to patients' health records to correctly estimate results of simulated activities (similarly to the critiquing mode of ProMA functioning described in [15]), which can have impact on future course of execution of the IIP. Moreover, after the IIP is created and stored in the system, the ProMA can support critiquing and verifying the execution of these procedures and alert health care specialist in charge in case of inconsistencies (see [15]).

6   Conclusions

In this paper we provided a detailed description of two multi-agent architectures and proposed their possible merge. First, we described the K4Care system – a knowledge-based multi-agent system supporting Home Care by means of eServices. Second, we presented a description of a general multi-agent architecture ProMA that can handle procedural knowledge captured in the form of processes. Finally we proposed an integration of both architectures by identifying the main similarities in both models and the main conceptual changes that needed to be done in order to successfully integrate these architectures on an implementation level. The most significant improvements for the K4Care system were noticeable in the process of creating a new Individual Intervention Plan for a patient. Finally, we identified possibilities for new features that the integration of ProMA into the K4Care system can further bring.

Acknowledgement. This research has been supported by the research program No. MSM 6840770012 "Transdisciplinary Research in Biomedical Engineering II" of the CTU in Prague.

References

1. Bosansky, B.: Process-based Multi-agent Architecture in Home Care. EJBI (1) (2010)
2. Gibert, K., Valls, A., Lhotska, L., Aubrecht, P.: Privacy Preserving and Use of Medical Information in a Multiagent System. In: Advances in Artificial Intelligence for Privacy Protection and Security - Intelligent Information Systems, vol. 1, pp. 165–193. World Scientific, London (2010)
3. Campana, F., et al.: Knowledge-Based HomeCare eServices for an Ageing Europe. K4CARE Deliverable D01, EU (2007)
4. Aubrecht, P., Matousek, K., Lhotska, L.: On designing EHCR Repository. In: HEALTHINF 2008 - International Conference on Health Informatics, pp. 280–285. IEEE, Piscataway (2008)
5. Riano, D.: The SDAModel: A Set Theory Approach. In: Proc. of the 20th IEEE Int. Symp. on Computer-Based Medical Systems, pp. 563–568 (2007)
6. Mendling, J.: Business process execution language for web services (BPEL). EMISA Forum 26(2), 5–8 (2006)
7. Scheer, W., Nuttgens, M.: ARIS architecture and reference models for business process management. In: Bus. Proc. Management, Models, Techniques, and Empirical Studies, pp. 376–389. Springer, London (2000)

8. Kumar, A., Smith, B., Pisanelli, M., Gangemi, A., Stefanelli, M.: Clinical guidelines as plans: An ontological theory. Methods of Information in Medicine 2 (2006)
9. Lenz, R., Reichert, M.: IT support for healthcare processes - premises, challenges, perspectives. Data Knowl. Eng. 61(1), 39–58 (2007)
10. Song, X., Hwong, B., Matos, G., Rudorfer, A., Nelson, C., Han, M., Girenkov, A.: Understanding requirements for computer-aided healthcare workflows: experiences and challenges. In: ICSE 2006: Proceedings of the 28th International Conference on Software Engineering, pp. 930–934. ACM, New York (2006)
11. Fox, J., Johns, N., Rahmanzadeh, A., Thomson, R.: Proforma: A method and language for specifying clinical guidelines and protocols, Amsterdam (1996)
12. Shahar, Y., Miksch, S., Johnson, P.: The asgaard project: a task specific framework for the application and critiquing of time-oriented clinical guidelines. Artificial Intelligence in Medicine 14, 29–51 (1998)
13. Peleg, M., Boxwala, A., Ogunyemi, O.: Glif3: The evolution of a guideline representation format. In: Proc. AMIA Annu. Fall Symp., pp. 645–649 (2000)
14. Isern, D., Moreno, A., Sanchez, D., Hajnal, A., Pedone, G., Varga, L.: Agent-based execution of personalised home care treatments. Applied Intelligence, 1–26 (2009)
15. Bosansky, B., Lhotska, L.: Agent-based process-critiquing decision support system. In: 2nd International Symposium on Applied Sciences in Biomedical and Communication Technologies, ISABEL 2009, pp. 1–6 (2009)

A Framework for the Production and Analysis of Hospital Quality Indicators

Alberto Freitas, Tiago Costa, Bernardo Marques, Juliano Gaspar,
Jorge Gomes, Fernando Lopes, and Isabel Lema

CINTESIS / CIDES, Faculty of Medicine,
University of Porto, Portugal
{alberto,tcosta,bmarques,jgaspar,jorge,
fernando,ilema}@med.up.pt

Abstract. Quality indicators are fundamental to health care managers as they can give valuable insight into how care is being delivered. Quality indicators are measures of health care quality that can make use of readily available hospital administrative data (e.g. inpatient data). This paper describes the development of a framework for the production and analysis of hospital quality indicators. The framework includes a set of national and international evaluated measures that can be calculated using already available data. The developed web-based framework is intuitive, user-friendly and is being continuously improved using users' feedback. It considers risk factors and allows comparing measures between time periods and also between hospitals and regions (benchmarking).

Keywords: quality indicators, performance indicators, administrative data, diagnosis related groups, benchmarking.

1 Introduction

1.1 Health Indicators

There is a general demand, with an increasing social pressure, for more information about the quality of healthcare. Information is important for patient's choice, for Government policy, and to alert and encourage providers and health professionals to improve quality of care. Performance indicators are useful to understand the quality of care provided, to compare health institutions and to highlight strengths and possible deficiencies. Under these circumstances, performance indicators are important tools for transparency, management and ultimately for improving the quality of care [1].

A performance indicator, also called quality indicator or management indicator, is usually a quantitative measure for some feature of an institution and can be used to screen, to compare and to evaluate the quality of a service [2]. It can be a rate or mean-based, with a quantitative basis for quality improvement, or it can be a sentinel,

C. Böhm et al. (Eds.): ITBAM 2011, LNCS 6865, pp. 96–105, 2011.
© Springer-Verlag Berlin Heidelberg 2011

identifying events of care that should trigger further investigation [3]. Indicators can be related to structure (e.g. material resources, human resources and organizational structure), process (e.g. appropriate therapy and patient's satisfaction), or outcome of health care (e.g. mortality and morbidity) [4].

In healthcare, clinical and performance indicators are elements that can be measured and can act as pointers to the quality of care, that is, they are not definitive measures [5]. Indicators can act as a flag for situations that might be wrong and should be further analysed, indicating either the presence or absence of potentially poor practices or outcomes. Indicators need to be reliable and valid, where reliability refers to consistency of measurement and validity refers to the degree to which an indicator measures what it is intended to measure.

It is yet important to avoid confusion between performance indicators and health outcomes [6]. While health outcomes are related to crude rates of adverse events in the population, performance indicators are related to characteristics of care that can be changed by the institutions (can be controlled by decision makers). Confounding factors, such as demographic factors, lifestyle factors, severity of illness, comorbidities, or the technical equipment of the hospital, can have an important impact on the outcome of care and therefore should be considered [3]. In this context, outcome measures must be adjusted for these factors, that is, risk adjustment is an essential procedure to induce a more fair and accurate inter-organizational (across hospitals or providers) comparison [7, 8].

In a planning phase, there are different steps required to develop and test indicators, including: choose the clinical area to evaluate (generic or disease-specific indicators), select and organize a measurement team, provide an overview of existing evidence (scientific literature) and practice, prioritize and select indicators, design measure specification (inclusion/exclusion criteria, risk adjustment, data sources), and perform pilot testing (reliability and validity) [9]. A good performance indicator should meet some important criteria. It should, for instance, reflect important features of health status, be sensitive to change, be based on reliable and valid information, be precisely defined, be easily quantifiable, and be relevant to policy and practice [10].

Performance indicators have been increasingly used in many countries to measure and improve the quality of hospitals [11-13]. They can be used for public reporting (to provide transparency in health care), quality improvement, comparisons (benchmarking/rating), pay-for-performance, or research [3, 14], and can also be of great relevance for the surveillance of health care quality [3]. In Portugal there is little available information about the use of performance indicators to support hospital management. Although, practice suggests that most hospitals calculate and monitor a range of performance measures.

1.2 Administrative Data

There are several data sources that can be used to measure performance, including administrative data (billing or claims data), electronic health records data, patient-derived data (questionnaires), reports and direct observation. These different data sources have different strengths and weaknesses.

Administrative data (AD) is routinely collected, widely available, relatively inexpensive, comprehends large amounts of data and, in general, covers all the country. Although with some data quality problems [15-18], AD is a valuable source for measuring quality of care. AD contains information form discharges, is used to bill and pay hospital services, has a standard format, can be used for many purposes, such as research or public reporting [19, 20]. AD is an important resource for hospital management and policy makers. In Portugal, as in many countries, AD typically contains demographic data (e.g., age, gender), "administrative data" (length of stay, type of admission, payer, discharge status) and ICD-9-CM codes for clinical data (diagnostics, procedures, external causes) [21].

1.3 Aim

The main aim of this project was to contribute to the improvement of quality in healthcare and management efficiency of hospitals through the use of performance indicators calculated from already available administrative central databases.

This project aimed to use secondary data to assess hospital performance through quality and performance indicators. Specifically, the project will contribute with the selection, integration and implementation of indicators based on available administrative data at the national level. The implemented tool should be used to increase the knowledge and improve the quality of management and hospital care through the analysis of local, regional and national quality and performance indicators.

2 Methods

The selection of indicators was a long and iterative process, and included a literature review (documentation and knowledge from scientific literature), the study of published hospital indicators, and interviews to common users (organizations and potential users of indicators). Special attention was given to the work of the Agency for Healthcare Research and Quality (AHRQ) [10, 11] and of the International Quality Indicator Project (IQIP) [12]. The expertise of hospital managers and specialists in clinical coding and auditing was also considered in all the process.

The implemented framework uses web-based technologies, specifically PHP, HTML e JavaScript. In this context, a user only needs a web browser to use the application.

We adopted a modular architecture that allows an easy and fast update of indicators, namely the software architecture Model-View-Controller (MVC) [13, 14].

For each indicator in the structure there is a *Model* component (files with extension php) a *View* component (files with extension tpl) and *Controller,* a component common to all indicators (file "indicators.php"). The configuration of this structure is saved in a database where it is possible, through a web interface, to change the configuration of any indicator when needed.

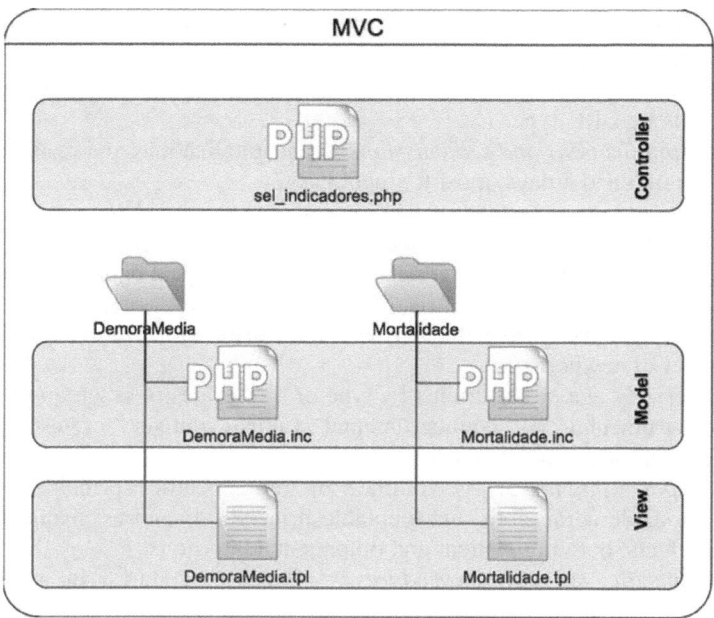

Fig. 1. File structure of the framework

After the definition of the structure, the following selected indicators were implemented and integrated in the framework:

- *Summary table* – a table for a quick view, summarizing the indicators, automatically highlighting some of the most important results.
- *Patients discharged* – number of patients discharged by type of production (emergent, planned, additional) and by destination after discharge.
- *Hospital mortality* – mortality within the hospital by age group, gender, Diagnosis Related Group (DRG), and principal diagnosis (only those with higher mortality rate).
- *Length-of-stay (LOS)* – global LOS, by type of admission, medical DRG, and surgical DRG.
- *Episodes with exceptional LOS* – episodes with short stay (those with LOS below DRG low trimpoint) and with long stay (those with LOS above DRG high trimpoint); ratio between patients discharged and equivalent patients; most common short and long stay DRG.
- *Invalid inpatient stays* – medical and surgical DRG, with ambulatory price comparison.
- *Long inpatient stays without comorbidities/complications* – most common DRG, diagnoses and procedures.
- *Surgical LOS* – global preoperative LOS, pre and postoperative LOS by DRG and type of production.
- *Not performed surgeries* – planned and non-planned surgeries, not performed, by cause (medical contraindication, decision of the patient, other causes).

- *Case-mix index* – global, inpatient stays, medical ambulatory, surgical ambulatory.
- *Equivalent patients* – relation between equivalent patients and discharged patients by DRG type.
- *Rehospitalizations and readmissions* – rehospitalizations and readmissions in a period of 3 and 5 days; most frequent DRG.
- *Outpatient care* – volume of ambulatory episodes by DRG type; medical and surgical ambulatory episodes, by type of production.
- *Inpatient vs. outpatient surgeries* – volume of surgical episodes in ambulatory and conventional surgery.
- *Newborns* – the proportion between newborns with problems and the total number of newborns.
- *Deliveries* – age of parturients by type of birth; rates of cesarean section; rates of instrumental vs. noninstrumental vaginal delivery; rates of epidural analgesia for vaginal delivery.
- *Principal diagnoses* – volume of non-specific principal diagnosis, questionable admissions, unacceptable principal diagnoses, manifestation and late effects, both in inpatient and outpatient admissions.
- *Non-specific surgical procedures* – main procedures in inpatient and outpatient admissions.
- *Complications in surgical DRG* – hemorrhages, lacerations, dehiscences, foreign bodies and postoperative infections in inpatient surgical DRG.
- *Complications in medical DRG* – pressure ulcers and urinary infections in inpatient medical DRG.
- *Specific pathologies* – Malignant neoplasms, strokes, and myocardial infarctions.
- *Inter-hospital transfers* – volume of inter-hospital transfers; comparison between the records of the original hospital and the records of the destination hospital.
- *DRG* – summary table with statistics for each DRG, namely total volume, total sum of hospitalization days, LOS, average age.

AHRQ Quality Indicators[1] are also being adapted and implemented in the framework. These quality indicators consist of four modules: Prevention Quality Indicators, Inpatient Quality Indicators, Patient Safety Indicators, and Pediatric Quality Indicators.

When considered relevant, indicators are being refined and adjusted to several factors (for instance, for age, case-mix, complexity of the hospital, DRG, comorbidities).

Another important indicator is related to data quality [18]. We developed a set of measures, consisting in more than one hundred rules, specifically for the detection of data quality problems.

To facilitate the selection, we created a page that allows the user to determine several conditions:

- The time period to consider for analysis (PT), consisting of an interval of dates, for instance those defining a calendar year.

---

[1] http://www.qualityindicators.ahrq.gov/

- The number of homologous periods to compare with PT.
- The number of contiguous periods to compare with PT.
- The set of hospitals or hospital centers to be analyzed, a combined analysis, a hospital separated analysis, or both.
- The set of indicators to be calculated and presented in the report.

Once these conditions are selected, the user can also decide between the report grouped by indicator or by institution (hospital). If selected by indicator", the report will be produced in order to show, for each indicator, all the selected hospitals. If, however, the option "by hospital" is selected, the resulting report will show, for each hospital, all the selected indicators. At the end, the obtained report can be printed or converted to a pdf format so it can easily be archived or sent via e-mail.

3   Results

The developed framework produces various and different types of tables for each indicator, resulting in more than 120 tables for one hospital, if all indicators are selected. Tables can be divided in three types: descriptive tables (Fig. 2), comparative tables (Fig. 3), and list of cases (Fig. 4 and 5).

22.1 - Inpatient DRG

MDC/DRG - Description	Total episodes		LOS sum		Length of stay (LOS)				Average age	Readm.		Rehosp.	
	N	% [1]	Σ	%	Min	Mean	Mdn	Max		N	%	N	%
MDC 0 - Pre- Major Diagnostic Categories													
103 HEART TRANSPLANT	6	0.01	414	0.12	32	69.00	60.50	115	49.83	0	0.00	0	0.00
302 KIDNEY TRANSPLANT	79	0.19	1,644	0.49	6	20.81	12.00	117	50.48	0	0.00	0	0.00
468 EXTENSIVE O.R. PROCEDURE UNRELATED TO PRINCIPAL DIAGNOSIS	135	0.32	2,099	0.62	0	15.55	9.00	273	48.52	8	0.64	7	0.62
470 UNGROUPABLE	13	0.03	104	0.03	1	8.00	6.00	24	40.08	4	0.32	3	0.26
477 NON-EXTENSIVE O.R. PROCEDURE UNRELATED TO PRINCIPAL DIAGNOSIS	69	0.16	860	0.25	1	12.46	6.00	74	45.97	2	0.16	2	0.18
482 TRACHEOSTOMY FOR FACE,MOUTH & NECK DIAGNOSES	34	0.08	630	0.19	1	18.53	11.00	79	61.38	0	0.00	0	0.00
483 ECMO OR TRACH W MECH VENT 96+ HRS OR TRACH W PDX EXCEPT FACE/MOUTH/NECK	182	0.43	13,743	~4.07	5	75.51	60.50	324	56.32	5	0.40	4	0.35
804 AUTOLOGOUS BONE MARROW TRANSPLANT	40	0.09	993	0.29	19	24.83	24.00	39	52.38	2	0.16	2	0.18
MDC 1 - Diseases and Disorders of the Nervous System													
1 CRANIOTOMY AGE >17 W CC	106	0.25	1,629	0.48	1	15.37	12.00	74	57.67	1	0.08	1	0.09
2 CRANIOTOMY AGE >17 W/O CC	171	0.40	1,989	0.59	1	11.63	10.00	84	56.92	1	0.08	1	0.09
6 CARPAL TUNNEL RELEASE	103	0.24	208	0.06	0	2.02	2.00	17	54.00	1	0.08	1	0.09

Fig. 2. Descriptive table for DRG, grouped by MDC

Generically, descriptive tables can be characterized as the ones composed by summary statistics. As an example, in the case of Fig. 2, we can see DRG grouped by Major Diagnostic Categories (MDC) and a set of measures used to summarize each DRG: average LOS, number of episodes, and average age.

Comparative tables are useful to analyze the hospital production in a given period and to easily compare with other chosen periods (homologous or contiguous). To facilitate the analysis, results can include images (e.g. arrows), automatically chosen considering the data, to highlight the differences in results of different time periods.

These images include a qualitative analysis for the tendency associated with the variation. Furthermore, each comparing period is associated with a different color in the table. As an example, tables presented in Fig. 3 show the evolution in the volume of discharged patients in each DRG type.

Tables with lists of cases are normally used to describe the most frequent diagnoses, procedures, and DRG in a specific indicator. We can see an example of this in Fig. 4 and 5 presenting, respectively, the ten DRG with most frequent hospital mortality and the 10 most frequent emergent admissions (ordered by average preoperative LOS).

1.1 - Discharged patients

Time period		DRG										
		Medical						Surgical					
		Urg	Prog	Add	Total	% (2)	Var (3)	Urg	Prog	Add	Total	% (2)	Var (3)
Analysis (12 months)	01/01/2008 to 31/12/2008	18,620	3,949	8	22,577	53.3		6,736	9,730	3,305	19,771	46.7	
Homologous annual comparison (1 years before)	01/01/2007 to 31/12/2007	19,080	3,852	12	22,944	55.7	1.60%	6,118	9,737	2,329	18,184	44.1	8.73%

1.10 - Discharged patients by discharge status

Time period			DRG						Others (1)		Total	
			Medical			Surgical						
			N	% (2)	Var (3)	N	% (2)	Var (3)	N	% (2)	N	% (4)
Analysis (12 months)	01/01/2008 to 31/12/2008	Home	19,638	51.9		18,216	48.1		13	0.0	37,867	89.4
		Another hospital	908	51.8		846	48.2		0	0.0	1,754	4.1
		Home health service	465	77.4		136	22.6		0	0.0	601	1.4
		Left against medical advice	369	77.8		105	22.2		0	0.0	474	1.1
		Expired	1,197	71.9		468	28.1		0	0.0	1,665	3.9
Homologous annual comparison (1 years before)	01/01/2007 to 31/12/2007	Home	19,407	53.7	1.19%	16,723	46.3	8.93%	2	0.0	36,132	87.7
		Another hospital	904	53.1	0.44%	716	42.0	18.16%	84	4.9	1,704	4.1
		Home health service	1,054	85.2	55.89%	183	14.8	25.69%	0	0.0	1,237	3.0
		Left against medical advice	379	80.6	2.64%	91	19.4	15.38%	0	0.0	470	1.1
		Expired	1,200	71.8	0.25%	471	28.2	0.64%	0	0.0	1,671	4.1

Fig. 3. Comparative tables for discharged patients

2.8 - Hospital mortality by principal diagnosis: Top 10 (age >65) - 01/01/2008 to 31/12/2008 - Analysis (12 months)

Code	Description	Deaths		Average age
		N	% (1)	
486	Pneumonia, organism unspecified	121	11.7	65.6
038.9	Unspecified septicemia	90	27.9	63.9
434.91	Cerebral artery occlusion, unspecified with cerebral infarction	65	7.7	71.4
431	Intracerebral hemorrhage	37	16.2	64.7
428.0	Congestive heart failure, unspecified	34	6.2	73.8
410.71	Subendocardial infarction, initial episode of care	25	5.5	67.7
507.0	Pneumonitis due to inhalation of food or vomitus	24	23.3	75.9
519.8	Other diseases of respiratory system, not elsewhere classified	23	7.4	66.1
162.3	Malignant neoplasm of upper lobe, bronchus or lung	19	16.4	64.1
162.8	Malignant neoplasm of other parts of bronchus or lung	18	19.4	62.7

Fig. 4. Table with the ten diagnoses with most frequent hospital mortality

7.2 - Most frequent emergency department-admitted DRG: Top 10 - 01/01/2008 to 31/12/2008

MDC	DRG	Description	No. of episodes	No. of days	Average LOS	Preoperative		Postoperative	
						Mean	% (1)	Mean	% (2)
10	287	SKIN GRAFT & WOUND DEBRID FOR ENDOC,NUTRIT & METAB DISORDERS	1	44	44.0	43.0	97.7	1.0	2.3
15	606	NEONATE, BIRTHWT 1000-1499G, W SIGNIF OR PROC, DISCHARGED ALIVE	1	69	69.0	41.0	59.4	28.0	40.6
0	103	HEART TRANSPLANT	4	316	79.0	29.3	37.0	49.8	63.0
24	701	HIV W O.R PROCEDURE & VENTILATOR OR NUTRITIONAL SUPPORT	6	310	51.7	23.2	44.8	28.5	55.2
17	874	LYMPHOMA & LEUKEMIA W MAJOR O.R. PROCEDURE W CC	2	78	39.0	22.5	57.7	16.5	42.3
24	700	TRACHEOSTOMY FOR HIV INFECTION	1	57	57.0	22.0	38.6	35.0	61.4
7	199	HEPATOBILIARY DIAGNOSTIC PROCEDURE FOR MALIGNANCY	1	24	24.0	18.0	75.0	6.0	25.0
8	866	LOCAL EXCISION & REMOVAL OF INT FIX DEVICES EXCEPT HIP & FEMUR W CC	3	109	36.3	17.7	48.6	18.3	50.5
24	702	HIV W O.R PROCEDURE W MULTIPLE MAJOR RELATED INFECTIONS	2	75	37.5	17.5	46.7	20.0	53.3
5	115	PRM CARD PACEM IMPL W AMI/HF/SHOCK OR AICD LEAD OR GNRTR PROC	1	24	24.0	17.0	70.8	7.0	29.2

Fig. 5. Table with the ten most frequent DRG in emergent production (ordered by preoperative LOS average)

4   Conclusion

Nowadays there is an increasing need to assess hospital production. This assessment allows managers to predict hospital activities and to manage resource allocation. In this setting, evidence-based information and tools are needed to help providers, managers, stakeholders, doctors, nurses and other hospital users to face this challenge. This project focus on these issues: information is needed for better decisions and to improve healthcare. Performance indicators should be reliable, timely, evidence-based, quantifiable, and based on relatively inexpensive administrative data. The main idea was achieved: we set up a web-based tool that can produce and compare hospital quality and performance indicators, and that allows each participant in hospital care, any time, any place, to be up to date, to be better informed about hospital activities and outcomes, and to be able to take timely decisions.

This framework is already being used both in local (hospitals) and in regional contexts (regional health administrations). The feedback received from users has been and is being fundamental for the continuous development and improvement of the framework. In fact, the feedback received reveals that the interactive development process, with the involvement of final users, produced a tool with many advantages when compared to other, more generic, tools.

The main characteristics of the tool include the easiness of use, the intuitive user interface, and mainly the way results are presented through well-defined and clearly organized tables, with different colorations. Results presented this way allow a simple and direct perception of the most relevant values, constituting an asset to management decision making and to the control of the efficiency of healthcare units.

Another advantage of this solution is that it can be easily used in many countries as the data model is mostly based in the so-called Minimum Basic Data Set (MBDS) [22]. The MBDS is common to many worldwide national health systems and, therefore, with modifications at the language level the tool can be easily configured (a translation to an English version is being done). The tool can also be easily fed with new data and can almost immediately, in real-time, allows an analysis of the evolution of indicators.

The implementation of AHRQ Quality Indicators (prevention, inpatient, patient safety indicators, and pediatric quality indicators) is an important add-on as it facilitates the comparison of indicators worldwide (because of its global, uniform, clearly defined methodology).

Other sets of specific quality indicators are also in study as, for instance, indicators of perinatal health and indicators for emergency care. These indicators will be produced using not only administrative databases (based on the MBDS) but also departmental databases.

The tool is and will be continuously improved. Special focus will be done to data quality (automatic detection and feedback of anomalies in data), to risk-adjustment, and to the improvement of the summary report, including intelligent and automatic analysis.

Case studies will emphasise the importance of using performance indicators. This tool will be used in the development of case studies to provide an in-depth analysis of disease-specific indicators (e.g. HIV/AIDS), to highlight the importance of using indicators in the study of ambulatorization, and to monitor the quality of services provided by hospitals through the use of a variable life-adjusted methodology [23].

Acknowledgments. The authors would like to acknowledge the research project HR-QoD - Quality of data (outliers, inconsistencies and errors) in hospital inpatient databases: methods and implications for data modeling, cleansing and analysis (project PTDC/SAU-ESA/75660/2006). The access to the data was provided by ACSS, I. P. (Administração Central do Sistema de Saúde, I. P.), the Portuguese Ministry of Health's Central Administration for the Health System.

References

1. Marshall, M.N., et al.: The public release of performance data: what do we expect to gain? A review of the evidence. JAMA 283(14), 1866–74 (2000)
2. Bourne, M., et al.: Designing, implementing and updating performance measurement systems. International Journal of Operations & Production Management 20(7), 754–771 (2000)
3. Mainz, J.: Defining and classifying clinical indicators for quality improvement. Int. J. Qual. Health Care 15(6), 523–530 (2003)
4. Donabedian, A.: Evaluating the quality of medical care (1966); Milbank, Q.: 83(4), 691–729 (2005)
5. Johantgen, M., et al.: Quality indicators using hospital discharge data: state and national applications. Jt. Comm. J. Qual. Improv. 24(2), 88–105 (1998)
6. Giuffrida, A., Gravelle, H., Roland, M.: Measuring quality of care with routine data: avoiding confusion between performance indicators and health outcomes. BMJ 319(7202), 94–98 (1999)
7. Iezzoni, L.I.: The risks of risk adjustment. JAMA 278(19), 1600–1607 (1997)
8. Shahian, D.M., et al.: Variability in the measurement of hospital-wide mortality rates. N. Engl. J. Med. 363(26), 2530–2539 (2010)

9. Mainz, J.: Developing evidence-based clinical indicators: a state of the art methods primer. Int. J. Qual. Health Care 15(suppl. 1), i5–i11 (2003)
10. Crampton, P., et al.: What makes a good performance indicator? Devising primary care performance indicators for New Zealand. NZ Med. J. 117(1191), U820 (2004)
11. AHRQ. Agency for Healthcare Research and Quality (February 2011), http://www.ahrq.gov/
12. CQC. Care Quality Commission (February 2011), http://www.cqc.org.uk
13. IQIP. International Quality Indicator Project (February 2011), http://www.internationalqip.com
14. Farquhar, M.: AHRQ Quality Indicators, in Patient Safety and Quality: An Evidence-Based Handbook for Nurses, R. Agency for Healthcare Research and Quality, MD (2008), http://www.ahrq.gov/qual/nurseshdbk/, Editor, AHRQ Publication No. 08-0043
15. Peabody, J.W., et al.: Assessing the accuracy of administrative data in health information systems. Med. Care 42(11), 1066–1072 (2004)
16. Szeto, H.C., et al.: Accuracy of computerized outpatient diagnoses in a Veterans Affairs general medicine clinic. Am. J. Manag. Care 8(1), 37–43 (2002)
17. Silva-Costa, T., et al.: Problemas de Qualidade de Dados em Internamentos Hospitalares e Possíveis Implicações. In: Saúde, R.A. (ed.) Sistemas e Tecnologias de Informação na, pp. 215–227. Edições Universidade Fernando Pessoa, Porto (2010)
18. Freitas, A., et al.: Implications of Data Quality Problems within Hospital Administrative Databases. In: XII Mediterranean Conference on Medical and Biological Engineering and Computing 2010, I. Proceedings, Editor, pp. 823–826. Springer, Heidelberg (2010)
19. Price, J., Estrada, C.A., Thompson, D.: Administrative data versus corrected administrative data. Am. J. Med. Qual. 18(1), 38–45 (2003)
20. Wright, J., et al.: Learning from death: a hospital mortality reduction programme. J. R. Soc. Med. 99(6), 303–308 (2006)
21. Iezzoni, L.I.: Assessing quality using administrative data. Ann. Intern. Med. 127(8 Pt. 2), 666–674 (1997)
22. Poss, J.W., et al.: A review of evidence on the reliability and validity of Minimum Data Set data. Healthc Manage Forum 21, 33–39 (2008)
23. Coory, M., Duckett, S., Sketcher-Baker, K.: Using control charts to monitor quality of hospital care with administrative data. International Journal for Quality in Health Care, 1–9 (2007)

Process Analysis and Reengineering in the Health Sector

Antonio Di Leva, Salvatore Femiano, and Luca Giovo

Università di Torino, Dipartimento di Informatica, corso Svizzera 185,
10149 Torino, Italy
{dileva,femiano,giovo}@di.unito.it

Abstract. In this paper we present the application of a methodology for business process analysis and reengineering to a care pathway for patients of an Oncology Hospital. This process has been analyzed to evaluate different drug administration modalities and the impact of these modalities to the organizational process.

Keywords: Methodology for Business Process Analysis, Process Simulation and Reengineering, Care Pathway.

1 Introduction

This paper presents the application of a methodology, called BP-M* (Business Process Methodology*), which is a practical approach for analysis, modeling, and reengineering of business processes [1]. It will be illustrated by means of the patient's care pathway in the Oncology Division of a large hospital in Torino (Italy).

BP-M* is a methodology which allows a precise definition of the patient care process and provide qualitative and quantitative information about the process, e.g.:

- the optimal type and number of resources (staff, rooms, beds, etc.),
- existing anomalies in the process (such as bottle-necks, long waiting times,…),
- suggestions to improve efficiency, *i.e.*, how to use resources in a better way, how to decrease patient length of stay in the department (cycle time),
- type of problems if something new happens (e.g. the workload increases).

It consists of four logically successive phases:

1. Context Analysis, that aims to fix the overall strategic scenario of the enterprise and to determine the organizational components which will be investigated;
2. Organizational Analysis and Process Engineering, which goal is to develop the so called As-Is model. This model provides managers and engineers with an accurate model of the enterprise as it stands, out of which they can make a good assessment of both available capabilities and the current status of business processes.
3. Process Diagnosis and Reorganization, that aims to specify the so called To-Be model, *i.e.* the set of restructured processes. The simulation approach helps to ensure that transformations applied to As-Is processes perform as required, and allows an effective "what-if" analysis.

C. Böhm et al. (Eds.): ITBAM 2011, LNCS 6865, pp. 106–107, 2011.

4. <u>Information System and Workflow Implementation</u>, which goal is to develop: a) the specification of the Information System environment, and b) the specification of the Workflow execution environment.

2   Case Study: The Patient Care Process in the Oncology Division

In this study we paid attention to the Out-Patients' Department of the Oncology Division where antiblastic therapies for the care of all solid tumors are administrated.

The process under analysis is the chemotherapy administration, i.e., the use of extremely powerful drugs to destroy cancer cells, which requires a very careful monitoring of patient conditions during drug infusion. The overall process is very complex and requires coordination among several doctors, nurses, pharmacists, laboratory and clerical personnel. Specifically, the intravenous administration of the Navelbina drug has been investigated and the patient cycle time (the overall time a patient spends in the department) is analyzed.

The As-Is model and two To-Be models have been developed, according to two different improvement scenarios. The first solution (sc.1) just changes the way to administrate the therapy (the oral administration of the drug has been introduced), with no changes in the department organization. The second solution (sc.2) is based on the first one, but introduces a set of technological innovations. By means of simulation of both the As-Is and two To-Be models, it is possible to evaluate some key performance indicators as <u>cycle time</u> (range of time that a patient spends in the Department) and <u>resource utilization</u> of the more critical resources (<u>doctors</u> and <u>nurses</u>). In our case, the following results have been obtained:

	As-Is	**To-Be (sc.1)**	**To-Be (sc.2)**
doctor %utilization	57%	47%	43%
nurse %utilization	53%	48%	42%
cycle time	205 min.	141 min.	107 min.

Based on these results, the Director of the Oncology Division approved the implementation of the first scenario. After a transitory period of time, experimental results were in good accordance with simulation outcome.

Reference

1. Di Leva, A., Femiano, S.: The BP-M* Methodology for Process Analysis in the Health Sector. Intelligent Information Management 3, 56–63 (2011)

Binary Classification Models Comparison: On the Similarity of Datasets and Confusion Matrix for Predictive Toxicology Applications

Mokhairi Makhtar, Daniel C. Neagu, and Mick J. Ridley

School of Computing, Informatics and Media, University of Bradford,
Bradford BD7 1DP, UK
{M.B.Makhtar,D.Neagu,M.J.Ridley}@Bradford.ac.uk

Abstract. Nowadays generating predictive models by applying machine learning and model ensembles techniques is a faster task facilitated by development of more user-friendly data mining tools. However, such progress raises the issues related to model management: once developed, many classifiers for example become accessible in collections of models. Choosing the relevant model from the collection can reduce costs of generating new predictive models: calculating the similarity of predictive models is the key to rank them, which may improve model selection or combination. For this aim we introduce a methodology to measure the similarity of classifiers by comparing their datasets, transfer functions and confusion matrices. We propose the Dataset Similarity Coefficient to calculate the similarity of datasets, and the Similarity of Models measure to calculate the similarity between such predictive models. In this paper we focus on toxicology applications of binary classification models. The results show that our methodology performs well in measuring models similarity from a collection of classifiers.

Keywords: Similarity of Predictive Models, Similarity of (Toxicology) Datasets, Confusion Matrix, Classifiers Comparison.

1 Introduction

The cycle of predictive models development includes data preparation, data quality check, reduction, modeling, prediction, and model performance analysis. Generating high-quality predictive models is a time consuming activity because of the tuning process in finding best model parameters. Some models are good for specific objectives, and some of them, in contrast, may be of average quality or even worse, while in other cases they may be the most appropriate models for the task at hand.

Predictive models acquire information through learning that can accurately recognize similar patterns. We define the predictive model structure consisting of *Input*, *(Transfer) Function* and *Output* [1]. The *Input* consists of data collections used by machine learning algorithms to get the prediction output (see Fig. 1). We define the *Output* using the model's confusion matrix, in the case of classification models. The *Transfer Function* is any type of machine learning algorithms used to generate

C. Böhm et al. (Eds.): ITBAM 2011, LNCS 6865, pp. 108–122, 2011.
© Springer-Verlag Berlin Heidelberg 2011

predictive classification models. The performance of the classification models is related to correctly classified instances. Such information can be found from the model's confusion matrix which is useful for classifier's performance evaluation.

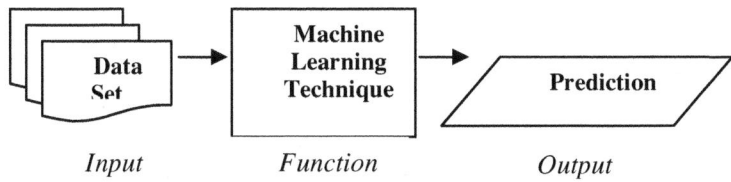

Input   *Function*   *Output*

Fig. 1. Predictive Modelling Framework

In this paper we propose a methodology to measure the similarity of classifiers as predictive models by comparing their input datasets and confusion matrices. In our work, we compare models built on similar input datasets which assure their compatibility for model selection and combination. The confusion matrix provides information on the performance of each class for a trained classifier. However, in order to measure the similarity of predictive models as a whole, we need to measure the similarity of predictive model elements (*Input*, *Function* and *Output*) independently. For the first structure element (*Input*) of the predictive model, we propose the *Dataset Similarity Coefficient (DSC)* to measure the similarity of two-dimensional datasets used to generate predictive models. The method based on *DSC* would measure the overall similarity of datasets between predictive models. We also propose the *Similarity of Predictive Model Function (SimF)* method to measure similarity of algorithms applied in generating the predictive models. For the last structure element (*Output*), we propose the *Similarity of Output (SimO)* method to measure the similarity of confusion matrices between compared predictive models. Performance measures such as Accuracy or False Negative Rate can be derived from the confusion matrix, which can also be used to derive further (customized) metrics.

Thus, to calculate the similarity of models, we combine similarity of predictive structures (DSC, SimF and SimO) together. We propose the *Similarity of Model (Sim)* method to measure similarity between predictive models. To give more flexibility to the users to calculate the similarity of predictive models, the *Sim* method allows the user to select which structure is important in this composite metric. Our proposed methodology is supported with detailed examples and explanations from predictive toxicology application. For this paper, we focus on toxicity prediction binary classification models. For the performance measures, we study the contribution of False Negative Rate (FNR) for predictive toxicology models, which is an important metric for the application domain: low value of the FNR means the predictive model is able to predict the toxicity of chemical compounds in a safer way.

The rest of the paper is structured as follows: Section 2 presents related work on binary similarity measures and predictive toxicology models comparison, and our motivation to compare binary classifiers with a composite similarity metric. Section 3 defines the technique of comparison of (toxicology) input datasets. The Similarity measure of Predictive Model (Transfer) Functions is proposed in Section 4. In Section

5 we introduce and exemplify the technique to compare the output of predictive models represented by their confusion matrix. A composite measure of the similarity of predictive models is proposed in Section 6. Experiments and results are discussed in Section 7. The paper ends with conclusions on current work and further directions. The results obtained for other toxicity datasets are shown in the Appendix.

2   Related Work

Comparison of predictive models can be accomplished by measuring the similarity between them. Similarity and distance metrics are complementary each other. For example the Hamming distance is one of the distances used to calculate the dissimilarity between two strings. The technique has successfully been applied to calculate the distance between features used in generating predictive models [1],[2].

Seung et. al [3] surveyed binary similarity and distance measures. They grouped the similarity and distance techniques into three main groups based on correlation, non-correlation and distance. Researchers can refer to either group for selecting the appropriate similarity measure to be applied, depending on the data. Lesot and Rifqi explored the similarity measures of different data types [4]. They found that the nature of data is the main factor deciding which similarity measure to be applied. Sequeira and Zaki [5] explored similarities across datasets using a two step solution: constructing a condensed model of the dataset and identifying similarities between the condensed models. Their technique is limited to finding similar subspaces based on the structure of the dataset alone, without sharing the datasets.

Comparison of predictive models is different compared to other similarity domains like sequence similarity in bioinformatics or information retrieval. Todeschini et. al [2] used variable cross-correlation matrix to find the relationship of features and reduce the similar features to find simpler models. They modified the Hamming distance technique to calculate the distances between predictive toxicology models.

In real cases, datasets are constantly updated. The changes in datasets will make previously generated predictive models obsolete, if not updated to the current content. This situation will have to be considered when comparing predictive models to calculate their similarity. The changes of instances could play a crucial role when finding similar models because it would affect the performances of learning models. Consequently the big challenge is to measure model similarity in a collection of models. There are four cases because of datasets update to be considered:

1. Different sets of records (instances) and similar variables (descriptors).
2. Different sets of records (instances) and different variables (descriptors).
3. Similar sets of records (instances) and similar variables (descriptors).
4. Similar sets of records (instances) and different variables (descriptors).

As far as we are aware, there are no comparisons of classifiers that calculate their similarity by incorporating the *Input*, *Function* and *Output* values in order to rank the classifiers from a collection of models. Many studies have been done by comparing the confusion matrices properties in ensembles of models [6], [7]. However, a more integrated approach to consider model development (training data and transfer function) is still necessary to improve model management and reuse for related tasks.

In this paper, we propose an improved version of predictive model comparison to accommodate the four cases of datasets changes and analyze their performance using the confusion matrix. It is an updated version of the method proposed earlier in [1] by adding new attributes of comparison which are related to the dataset. The number of instances and the number of common attributes in both predictive models compared is an important measure to calculate the similarity of models. This factor will help in measuring similarity between models.

3   Similarity of Toxicology Datasets

In this section we introduce the technique to compare datasets used for development of predictive models from a collection of models. The models are generated using Weka (see [8]) and are based on the four cases introduced above. The aim is to analyze if similar models would predict similar results.

In our case, the datasets are composed of rows (chemical compounds) and columns (descriptors): the descriptors are calculated values to describe the chemical compound properties, whereas the outputs are toxicity values obtained from testing chemical compounds against in vivo or in vitro end points. Simpler predictive models can be generated if followed a feature selection process, which is applied to find the most relevant descriptors of the dataset.

For the evaluation of the similarity between the two sets, we can use Jaccard Similarity Coefficient (JSC), as the measure defined as the size of the intersection divided by the size of the union of the sets A and B:

$$JSC(A,B) = \frac{|A \cap B|}{|A \cup B|} \tag{1}$$

where $A = \{a_i\}$ and $B = \{b_i\}$, containing $i = 1..n$ tuples.

In our case, the datasets are two-dimensional sets (see Table 1 to Table 4 for examples), and thus cannot be measured using JSC. This study addresses the cases where the values for the same descriptors and chemical compounds are the same (data quality check have been previously done) for all the datasets experimented.

We propose a new method called Dataset Similarity Coefficient (DSC) to measure the similarity of two-dimensional datasets used to generate predictive models. The DSC between Model A (M_a) and Model B (M_b) datasets is:

$$DSC_{(Ma,Mb)} = \frac{|C_{Ma} \cap C_{Mb}||R_{Ma} \cap R_{Mb}|}{|C_{Ma} \cup C_{Mb}||R_{Ma} \cup R_{Mb}|} \tag{2}$$

C_{Ma} is the set of all descriptor names (columns) for the dataset used in Model A,
C_{Mb} is the set of all descriptor names (columns) for the dataset used in Model B,
R_{Ma} is the set of all chemical compound attributes (rows) for the dataset used in M_a,
R_{Mb} is the set of all chemical compound attributes (rows) for the dataset used in M_b.
From Equation 2, $DSC_{(Ma, Mb)}$ is a Dataset Similarity Coefficient used to measure similarity of *Input* dataset for Model A (M_a) and Model B (M_b).

We will exemplify the use of our proposed DSC: let us say we have four models with their particular datasets which are DS_1, DS_2, DS_3 and DS_4 (see Table 1 to Table 4). Dataset DS_1 is the main dataset whereas the other datasets are subsets of DS_1. There are seven descriptors (ID, D1, D2, D3, D4, D5 and Class) in these datasets.

Table 1. Dataset DS_1

ID	D_1	D_2	D_3	D_4	D_5	Class
1	9.5	7.5	10	7.5	21.5	Yes
2	1.1	2	4.1	10	20	No
3	7	10	11	10.6	20.5	Yes
4	10	15	20	15	20	Yes
5	9	14	19	14	10	No

Table 2. Dataset DS_2

ID	D_1	D_2	Class
1	9.5	7.5	Yes
2	1.1	2	No
3	7	10	Yes

Table 3. Dataset DS_3

ID	D_2	D_3	Class
3	10	11	Yes
4	15	20	Yes
5	14	19	No

Table 4. Dataset DS_4

ID	D_4	D_5	Class
2	10	20	No
3	10.6	20.5	Yes
4	15	20	Yes

Table 5 shows the similarity coefficient matrix of datasets DS_1, DS_2, DS_3 and DS_4 calculated by using Dataset Similarity Coefficient: DS_2 seems to be most similar to DS_1, followed by DS_3 and DS_4. For example, the similarity between Model A (M_a) and Model B (M_b) that used datasets DS_1 and DS_2 is: $DSC_{(Ma,Mb)}=(|3||3|)/(|6||5|)=9/30=0.30$. DSC may provide an effective measure in calculating the similarity of datasets used in predictive models. The dataset similarity for two predictive models will give us an indication of what the predictive model may come from similar datasets.

Table 5. Dataset similarity coefficient matrix of dataset DS_1, DS_2, DS_3 and DS_4

	DS_1	DS_2	DS_3	DS_4
DS_1	1	0.30	0.30	0.30
DS_2	0.30	1	0.10	0.05
DS_3	0.30	0.10	1	0.05
DS_4	0.30	0.05	0.05	1

4 Similarity of Predictive Model Functions

In this section, we introduce the similarity measure for the second element of the predictive model, the *Transfer Function F*. We propose to apply the Jaccard Similarity Coefficient (JSC) to calculate the similarity for *F*; it is defined as the size of the intersection divided by the size of the union of the set F_{Ma} and set F_{Mb}:

$$SimF_{(Ma,Mb)} = \frac{\left|F_{Ma} \cap F_{Mb}\right|}{\left|F_{Ma} \cup F_{Mb}\right|} \tag{3}$$

For consistency we assume all parameter names of predictive models come from the same representation such as Predictive Toxicology Markup Language (PTML) [1] or Predictive Model Markup Language (PMML) [9].

For example, given two models Model A (M_a) and Model B (M_b):

F_{Ma} = {"DecisonTree", "10-Folds, "classification"} and
F_{Mb} = {"NeuralNet", "10-Folds","classification"},

the similarity of the two transfer functions is $SimF_{(Ma,Mb)}$=2/3=0.67. The result for $SimF_{(Ma, Mb)}$ shows 67% intersection between the function sets of the two models.

5 Similarity of Confusion Matrices

The predictive models comparison helps finding how similar models are. Relying only on standard performance indicators such as accuracy may not give much clue on the quality of a predictive model. Sometimes the accuracy might be biased for certain class and this will not provide good indications of predictive model's performance.

In this paper, we propose a novel technique to compare predictive models performance based on their confusion matrices. The confusion matrix stresses the raw results of the classification generated by the classification algorithm. The result contains information on correct and incorrect classification determined by the machine learning algorithm to predict the output.

A confusion matrix [10] contains information about actual and predicted classifications done by a classification model. Performance of such models is commonly evaluated using the data in the matrix (see Table 6). Table 6 shows the confusion matrix for a two class classifier.

Table 6. Confusion Matrix of Binary Classification: True Positive (TP), True Negative (TN), False Negative (FN) and False Positive (FP)

			Actual		
			Positive	Negative	
			Yes	No	
Predicted		Positive	**(TP)**	**(FP)**	*Yes*
		Negative	**(FN)**	**(TN)**	*No*

The performance measures can be calculated as follows [10], [11]:

TPRate = TP/(TP+FN) is the number of correct predictions for positive output (e.g. *Yes*), *FPRate = FP/(FP+TN)* is the number of incorrect predictions for the negative output (e.g. *No*), *FNRate = 1-TPRate* is the number of incorrect prediction for the positive output , and *TNRate =1-FPRate* is the number of correct predictions for the negative output. *Accuracy = (TP+TN)/(TP+FP+FN+TN)* is the number of correct predictions for all classes.

Table 7. Confusion Matrix for Model M1, Model M2 and Model M3

		Model M1			Model M2			Model M3	
		Actual			Actual			Actual	
		Yes	No		Yes	No		Yes	No
Predicted	Yes	1	2	Yes	3	4	Yes	2	3
	No	3	4	No	1	2	No	2	3

For example, let's consider the confusion matrices for three models M1, M2 and M3 (shown in Table 7) with the same binary output classes from the same input dataset. The following confusion matrices resulted from the classifiers learning whether a chemical compound is toxic (class "*Yes*") or non-toxic (class "*No*"). All models show the same accuracy value (see Table 8) although the confusion matrices are different. The first classifier (Model M1 in Table 7) successfully classifies 5 out of 10 cases. However, an alarming 3 chemical compounds will be given the all clear when they are actually toxic. Also the 2 chemical compounds said to be toxic despite they are not will be rejected although incorrectly.

Table 7 also shows the confusion matrix for the second classifier (Model M2). This time the model classifies well the class "*Yes*" but worse the class "*No*". Overall, it correctly classifies 50% of all cases and shows a very different confusion matrix than Model M1. The confusion matrix of the model M3 shows a more balanced behavior than first two classifiers, according to the TP, FP, FN and TN values. However its accuracy is the same as of M1 and M2. This shows that comparing models may require a more detailed and composite performance measure, since accuracy alone does not define fully the predictive models performance.

The confusion matrices help to evaluate classifiers in a more detailed way than just using the accuracy score and also can provide a tool to compare models' performance. Below we propose a methodology to compare confusion matrices. Table 8 contains *Accuracy, TPR, TNR, FNR* and *FPR* values for models M1, M2 and M3, calculated from their confusion matrices: although accuracy is the same for all models, it fails to describe differences in their performance. The other four performance indicators (*TPR, TNR, FNR* and *FPR*) help providing detailed performance for each class and are more realistic tools for comparing the performance of the predictive models.

Table 8. Performance Measures (*Accuracy, TPR, TNR, FNR and FPR*) for Models M1, M2 and M3

Models	Accuracy	TPR	TNR	FNR	FPR
M1	0.5	0.25	0.67	0.75	0.33
M2	0.5	0.75	0.33	0.25	0.67
M3	0.5	0.5	0.5	0.5	0.5

The Euclidean Distance can be used to calculate the difference between performances of the two models. In this example we chose *TPR* and *TNR* to measure the distance between the models performances. For the performance measures in Table 8, let's use the notations $k_1, ... k_n$. In this case k_1 is *TPR*, k_2 is *TNR*, where *n* equals 2. The following steps illustrate the calculation of the distance between the confusion matrices between two predictive models.

Step 1: Save the selected performance measure/s in a 1-dimension (vector) format.
We save the selected performance measures into two rows vectors. The vectors of performance measures for M1 and M2 where k_1 is *TPR* and k_2 is *TNR* are:
V_{Ma} = (0.25, 0.67) and V_{Mb} = (0.75, 0.33). From the vectors V_{Ma} and V_{Mb}, we can calculate the distance between them by using the distance technique.

Step 2: Calculate the distance between the vectors.
The distance is calculated using the Euclidean Distance:

$$d_{ij} = \sqrt{\sum_{k=1}^{n}(x_{ik} - x_{jk})^2} \qquad (4)$$

The distance *O* (Output) between Model A(M_a) and Model B (M_b) is the average of distances between the confusion matrix elements;

$$DistO_{(Ma,Mb)} = \frac{1}{n}\sqrt{\sum_{k=1}^{n}(V_{Mak} - V_{Mbk})^2} \qquad (5)$$

Similarity and distance measures are complementing any one to each other. In our case, the similarity of output *O (SimO)* between two models will be:

$$SimO_{(Ma,Mb)} = 1 - \frac{1}{n}\sqrt{\sum_{k=1}^{n}(V_{Mak} - V_{Mbk})^2} \qquad (6)$$

where: *k* is the order of performance measures selected, *n* equals to number of *k*, $V_{(Ma)}$ is the vector for Model A (M_a), and $V_{(Mb)}$ is the vector for Model B (M_a).
The value for $SimO_{(Ma,Mb)}$ in the example above is 0.75. Table 9 contains the values for $SimO_{(Ma,Mb)}$ related to the similarity of the three classifiers using *TPR* and *TNR*. The models might have learned identical data sets but were generated using different classification algorithms.

Table 9. Similarity Matrix For Model M1, M2 and M3

	M1	M2	M3
M1	1	0.42	0.21
M2	0.42	1	0.21
M3	0.21	0.21	1

6   Similarity of Predictive Models

In the previous sections, we proposed to evaluate the similarity of a predictive model based on the similarity of input (I) datasets, the performance measured by the Confusion Matrix as the output (O), and the similarity of the transfer function/s. Let's say we have the similarity value for I (DSC), F and O ($SimO$) for models M1, M2 and M3 (see Table 10):

Table 10. Value of I, F and O for Model M1, M2 and M3

	M1	M2	M3
M1	-	$I=0, F=1, O=0.3$	$I=0.9, F=0.2, O=0.2$
M2	$I=0, F=1, O=0.3$	-	$I=1, F=0.9, O=0.7$
M3	$I=0.9, F=0.2, O=0.2$	$I=1, F=0.9, O=0.7$	-

To find the similarity between models, we propose to combine all similarity values for *Input* (I), *Function* (F) and *Output* (O) according to our definition of predictive models structure. To provide more flexibility in calculating the similarity of predictive models, each structure of a predictive model has its own weight α, β, γ. We propose the Similarity of Models:

$$Sim_{(Ma,Mb)} = \frac{\alpha \times I_{(Ma,Mb)} + \beta \times F_{(Ma,Mb)} + \gamma \times O_{(Ma,Mb)}}{\alpha + \beta + \gamma} \tag{7}$$

where $I_{(Ma,Mb)}$ is the Dataset Similarity Coefficient ($DSC_{(Ma,Mb)}$) between Model A (M_a) and Model B (M_b), $F_{(Ma,Mb)}$ is the Similarity of Function ($SimF_{(Ma,Mb)}$) between Model A (M_a) and Model B (M_b), $O_{(Ma,Mb)}$ is the Similarity of Confusion Matrix ($SimO_{(Ma,Mb)}$) between Model A (M_a) and Model B (M_b), and $\alpha, \beta, \gamma \in [0,1]$ are real numbers.

We can handle the values of α, β or γ depending on the priority given to the predictive model's elements. Let's say we assign the weight value for I ($\alpha=1$), O ($\gamma=1$) and F ($\beta=0$). The similarity ($Sim_{(Ma,Mb)}$) between models is shown in Table 11 from where we learn that M1 and M2 are less similar than M2 is to M3.

Table 11. Similarity values of Models M1, M2, M3 given I ($\alpha=1$), O ($\gamma=1$) and F ($\beta=0$)

	M1	M2	M3
M1	1	0.15	0.55
M2	0.15	1	0.85
M3	0.55	0.85	1

7   Experiments and Results

For this study, we generated collections of models using a series of algorithms implemented in Weka [8], such as k-nearest neighbors classifier (weka.classifiers.lazy.IBk) and decision trees (weka.classifiers.trees.J48). The predictive models were applied to various toxicology data sets such as the Demetra [12] and TETRATOX [13]. Every dataset had originally more than two classes to predict the toxicology levels for every compound. Since the objective in this paper is the study of the similarity of predictive binary classifiers and their performance, we mapped the old classes onto new binary classes (see Table 12).

We generated 386 predictive models with different combinations of datasets, algorithms, and model parameters. The models were generated from a group of predictive toxicology data sets whereby each group of data set was run through data preparation and reductions processes. The data sets had been split into 70% (training data set) and 30% (testing data set). We also examined the relationships of the datasets. The Correlation-based Feature Selection (CFS) algorithm was applied to the datasets to find sets of attributes that are highly correlated with the target classes [14]. Each data set was run using Weka with 10-fold cross validation and classifiers weka.classifiers.lazy.IBk and weka.classifiers.trees.J48.

The datasets for each end point were split into three categories which are 100% of the instances, 70% of the instances for training and 30% of the instances for testing. The last two sets are subsets of the original set and they are both different to each other. For each dataset, descriptors were either selected using a feature selection algorithm (CFS) or not (None), and we generated models using machine leaning algorithms from Weka (IBK or J48).

Experiment 1:

The similarity of predictive models ($Sim_{(Ma,Mb)}$) was calculated in this experiment for I (α=1), O (γ=1) and F (β=0). False Negative Rate (*FNR*) was set in the ($SimO_{(Ma,Mb)}$) to justify the importance of it from the viewpoint of toxicology datasets, where we want the model to have low *FNR*. This means that the models were chosen if having less FN rate. For example in Table 13, similar datasets are likely to predict with similar *FNR* although using different machine learning algorithms, such as Model 1 and Model 151, and Model 4 and Model 154. In Table 16 to Table 19 (see the Appendix) similar results obtained for other toxicity datasets (Daphnia, Dietary Quail, Oral Quail and Trout) are shown.

Table 12. The mapping of the old classes to the new binary classes in each datasets

Data sets	Old Classes	New Classes (2-Classes)	Instances
Trout	Class1, Class2	Yes (Toxic)	218
	Class3	No (Non-toxic)	64
Oral Quail	Class1, Class2,Class3	Yes (Toxic)	56
	Class4	No (Non-toxic)	60
Daphnia	Class1, Class2	Yes (Toxic)	187
	Class3,Class4	No (Non-toxic)	77
Dietary Quail	Class1, Class2,Class3	Yes (Toxic)	101
	Class4,Class5	No (Non-toxic)	22
Bee	Class1, Class2,Class3,Class4	Yes (Toxic)	76
	Class5	No (Non-toxic)	29

Table 13. Results of model similarity from Bee dataset

Machine Learning	IBK						J48					
Feature Selection	None			CFS			None			CFS		
Split (%)	100	70	30	100	70	30	100	70	30	100	70	30
Model	**1**	**2**	**3**	**4**	**5**	**6**	**151**	**152**	**153**	**154**	**155**	**156**
1	1	0.84	0.62	0.46	0.48	0.44	0.97	0.84	0.64	0.45	0.46	0.49
2	0.84	1	0.48	0.45	0.47	0.43	0.8	0.99	0.5	0.44	0.44	0.48
3	0.62	0.48	1	0.43	0.45	0.41	0.59	0.47	0.98	0.42	0.43	0.46
4	0.46	0.45	0.43	1	0.83	0.63	0.49	0.46	0.45	0.99	0.84	0.62
5	0.48	0.47	0.45	0.83	1	0.46	0.49	0.48	0.47	0.82	0.98	0.49
6	0.44	0.43	0.41	0.63	0.46	1	0.47	0.44	0.43	0.64	0.48	0.95
151	0.97	0.8	0.59	0.49	0.49	0.47	1	0.81	0.61	0.48	0.49	0.47
152	0.84	0.99	0.47	0.46	0.48	0.44	0.81	1	0.49	0.45	0.45	0.49
153	0.64	0.5	0.98	0.45	0.47	0.43	0.61	0.49	1	0.44	0.44	0.48
154	0.45	0.44	0.42	0.99	0.82	0.64	0.48	0.45	0.44	1	0.84	0.61
155	0.46	0.44	0.43	0.84	0.98	0.48	0.49	0.45	0.44	0.84	1	0.46
156	0.49	0.48	0.46	0.62	0.49	0.95	0.47	0.49	0.48	0.61	0.46	1

Experiment 2:

The objective of the second experiment is to find the similarity of datasets between five end points. The five Demetra datasets are Bee, Daphnia, Dietary Quail, Oral Quail and Trout [12]. For this experiment, I (α =1), O (γ =0) and F (β =0). From the result (see Table 14), all datasets share over 50% similar descriptors and chemical compounds: the highest dataset similarity is 63% between Daphnia and Trout, while Bee and Oral quail have about 48% chemical compounds in common.

Table 14. Results of similarity for all datasets

Data Sets	Bee	Daphnia	Dietary Quail	Oral Quail	Trout
Bee	1	0.54	0.59	0.48	0.58
Daphnia	0.54	1	0.59	0.53	0.63
Dietary Qual	0.59	0.59	1	0.56	0.59
Oral Quail	0.48	0.53	0.56	1.01	0.5
Trout	0.58	0.63	0.59	0.5	1

Experiment 3:

This experiment is designed to show that the performance of models rely on the functions used to generate the predictive models. In this experiment we want to compare the results if the models used feature selection algorithms with different classifiers (see Table 15).

Table 15. Results of the Accuracy and the False Negative Rate (FNR) for all datasets

Dataset	Bee		Daphnia		Dietary Quail		Oral Quail		Trout	
Feature Selection	None	CFS	None	CFS	None	CFS	None	CFS	None	CFS
IBK										
Model	**M1**	**M4**	**M31**	**M34**	**M61**	**M64**	**M91**	**M94**	**M121**	**M124**
Acc	82.85	88.57	73.11	74.62	76.42	81.30	67.24	63.79	78.72	80.49
FNR	0.12	0.04	0.19	0.20	0.19	0.15	0.32	0.38	0.14	0.12
J48										
Model	**M151**	**M154**	**M181**	**M184**	**M211**	**M214**	**M241**	**M244**	**M271**	**M274**
Acc	86.67	89.52	74.62	78.41	74.80	82.11	67.24	73.28	74.11	81.92
FNR	0.06	0.02	0.19	0.12	0.20	0.13	0.36	0.30	0.17	0.07

From Table 15, generally the accuracy of the models increased when a feature selection algorithm was used. In this experiment we used the Correlation-based Feature Selection (CFS) as feature selection algorithm. From the toxicology point of view, we are interested in the *FNR*, whether the models are able to minimize the error prediction of the toxic class accurately or not. From the results, models with feature selection and using J48 classifier seem to be a good combination in correctly predicting the toxicity class.

8   Conclusions

This study shows that comparing predictive models is an important issue since it can help users minimizing the cost of generating new predictive models by reusing existing ones. The comparison of models from huge repositories of models would help to find the relevant and good models based on comparison algorithms. The confusion matrix provides a more useful quality indicator for the performances of predictive classifier models. The analysis and understanding of their relationships will make the classifier selection more reliable.

The comparison of models from large repositories of models would help to find the relevant model based on optimization of comparison functions. Our experiments also show that the similarity of models will help classifying models for further analysis and customized selection and combination of the relevant model according to the user's needs. In the future, we want also to consider the datasets from other predictive toxicology sources and improve the comparison of models' transfer function properties by updating the weights in order to make the similarity *Sim* definition adaptive to optimization of ranking the classifiers from collection of models. Further work will also address the automation of quality check on similar algorithms with different parameters or different training algorithms.

Acknowledgments. This work is partially supported by BBSRC, TSB and Syngenta through the Knowledge Transfer Partnerships (KTP) Grant "Data and Model Governance with Applications in Predictive Toxicology". The first author acknowledges the financial supports received from the University Sultan Zainal Abidin (UniSZA), Malaysia.

References

[1] Makhtar, M., Neagu, D.C., Ridley, M.: Predictive Model Representation and Comparison: Towards Data and Predictive Models Governance. In: Proceedings of the 10th UK Workshop on Computational Intelligence UKCI 2010, pp. 1–6. University of Essex, UK (2010)

[2] Todeschini, R., Consonnia, V., Pavan, M.: A distance measure between models: a tool for similarity/diversity analysis of model populations. Chemometrics and Intelligent Laboratory Systems 70, 55–61 (2004)

[3] Choi, S.-S., Cha, S.-H., Tappert, C.C.: A Survey of Binary Similarity and Distance Measures. Journal of Systemics, Cybernetics and Informatics 8, 43–48 (2010)

[4] Lesot, M.-J., Rifqi, M.: Similarity measures for binary and numerical data: a survey. International Journal of Knowledge Engineering and Soft Data Paradigms 1, 63–84 (2009)

[5] Sequeira, K., Zaki, M.J.: Exploring Similarities across High-dimensional Datasets. In: Taniar, D. (ed.) Research and Trends in Data Mining Technologies and Applications, vol. 3, pp. 53–85. Idea Group Inc., USA (2007)

[6] Prasanna, S.R.M., Yegnanarayana, B., Pinto, J.P., Hermansky, H.: Analysis of Confusion Matrix to Combine Evidence for Phoneme Recognition. IDIAP Research Report, IDIAP-RR-27-2007 (2007)

[7] Freitas, C.O.A., Carvalho, J.M.D.: J. Jose Josemar Oliveira, S. B. K. Aires, and R. Sabourin.: Confusion Matrix Disagreement for Multiple Classifiers. In: Proceedings of the Congress on pattern recognition 12th Iberoamerican Conference on Progress in Pattern Recognition, Image Analysis and Applications, pp. 387–396 (2007)

[8] Witten, I.H., Frank, E., Trigg, L., Hall, M., Holmes, G., Cunningham, S.J.: Weka: Practical Machine Learning Tools and Techniques with Java Implementations. In: Proceedings of the ICONIP/ANZIIS/ANNES 1999 Workshop on Emerging Knowledge Engineering and Connectionist-Based Information Systems, pp. 192–196 (1999)

[9] D. M. Group.: PMML 3.2 - Model Explanation Documents (2008)

[10] Kohavi, R., Provost, F.: Glossary of Terms. Editorial for the Special Issue on Applications of Machine Learning and the Knowledge Discovery Process 30, 271–274 (1998)

[11] Fawcett, T.: ROC Graphs: Notes and Practical Considerations for Researchers. HP Laboratories (2004)

[12] DEMETRA Project (2008), http://www.demetra-tox.net/

[13] TETRATOX.: TETRATOX Home (2008),
http://www.vet.utk.edu/TETRATOX/index.php

[14] Trundle, P.: Hybrid Intelligent Systems Applied to Predict Pesticides Toxicity - a Data Integration Approach. Phd Thesis. School of Informatics, University of Bradford, UK (2008)

Appendix

Table 16. Results of model similarity from Daphnia dataset

Machine Learning	IBK						J48					
Feature Selection	None			CFS			None			CFS		
Split (%)	100	70	30	100	70	30	100	70	30	100	70	30
Model	**31**	**32**	**33**	**34**	**35**	**36**	**181**	**182**	**183**	**184**	**185**	**186**
31	1	0.84	0.61	0.5	0.5	0.49	1	0.83	0.62	0.46	0.45	0.5
32	0.84	1	0.44	0.48	0.49	0.49	0.84	1	0.45	0.48	0.47	0.48
33	0.61	0.44	1	0.46	0.45	0.45	0.61	0.44	0.99	0.42	0.41	0.46
34	0.5	0.48	0.46	1	0.84	0.64	0.49	0.48	0.47	0.96	0.8	0.65
35	0.5	0.49	0.45	0.84	1	0.5	0.5	0.49	0.46	0.81	0.96	0.5
36	0.49	0.49	0.45	0.64	0.5	1	0.5	0.49	0.46	0.62	0.46	0.99
181	1	0.84	0.61	0.49	0.5	0.5	1	0.84	0.62	0.47	0.46	0.49
182	0.83	1	0.44	0.48	0.49	0.49	0.84	1	0.45	0.48	0.47	0.48
183	0.62	0.45	0.99	0.47	0.46	0.46	0.62	0.45	1	0.43	0.42	0.47
184	0.46	0.48	0.42	0.96	0.81	0.62	0.47	0.48	0.43	1	0.84	0.61
185	0.45	0.47	0.41	0.8	0.96	0.46	0.46	0.47	0.42	0.84	1	0.45
186	0.5	0.48	0.46	0.65	0.5	0.99	0.49	0.48	0.47	0.61	0.45	1

Table 17. Results of model similarity from Dietary Quail dataset

Machine Learning	IBK						J48					
Feature Selection	None			CFS			None			CFS		
Split (%)	100	70	30	100	70	30	100	70	30	100	70	30
Model	**61**	**62**	**63**	**64**	**65**	**66**	**211**	**212**	**213**	**214**	**215**	**216**
61	1	0.82	0.62	0.48	0.49	0.47	0.99	0.85	0.64	0.47	0.5	0.47
62	0.82	1	0.44	0.45	0.46	0.44	0.83	0.97	0.46	0.44	0.47	0.44
63	0.62	0.44	1	0.49	0.48	0.5	0.61	0.47	0.98	0.5	0.47	0.5
64	0.48	0.45	0.49	1	0.85	0.64	0.47	0.48	0.49	0.99	0.83	0.64
65	0.49	0.46	0.48	0.85	1	0.48	0.48	0.48	0.49	0.83	0.98	0.48
66	0.47	0.44	0.5	0.64	0.48	1	0.46	0.47	0.48	0.65	0.47	1
211	0.99	0.83	0.61	0.47	0.48	0.46	1	0.85	0.64	0.46	0.5	0.46
212	0.85	0.97	0.47	0.48	0.48	0.47	0.85	1	0.49	0.47	0.5	0.47
213	0.64	0.46	0.98	0.49	0.49	0.48	0.64	0.49	1	0.48	0.49	0.48
214	0.47	0.44	0.5	0.99	0.83	0.65	0.46	0.47	0.48	1	0.82	0.65
215	0.5	0.47	0.47	0.83	0.98	0.47	0.5	0.5	0.49	0.82	1	0.47
216	0.47	0.44	0.5	0.64	0.48	1	0.46	0.47	0.48	0.65	0.47	1

Table 18. Results of model similarity from Oral Quail dataset

Machine Learning	IBK						J48					
Feature Selection	None			CFS			None			CFS		
Split (%)	100	70	30	100	70	30	100	70	30	100	70	30
Model	91	92	93	94	95	96	241	242	243	244	245	246
91	1	0.83	0.62	0.47	0.47	0.36	0.98	0.84	0.64	0.49	0.44	0.4
92	0.83	1	0.45	0.46	0.49	0.34	0.82	0.97	0.48	0.49	0.46	0.4
93	0.62	0.45	1	0.49	0.44	0.39	0.63	0.48	0.97	0.46	0.41	0.5
94	0.47	0.46	0.49	1	0.79	0.53	0.49	0.48	0.48	0.96	0.77	0.6
95	0.47	0.49	0.44	0.79	1	0.33	0.45	0.46	0.46	0.83	0.97	0.4
96	0.36	0.34	0.39	0.53	0.33	1	0.37	0.37	0.36	0.5	0.3	0.9
241	0.98	0.82	0.63	0.49	0.45	0.37	1	0.84	0.64	0.47	0.43	0.5
242	0.84	0.97	0.48	0.48	0.46	0.37	0.84	1	0.5	0.48	0.43	0.5
243	0.64	0.48	0.97	0.48	0.46	0.36	0.64	0.5	1	0.49	0.44	0.4
244	0.49	0.49	0.46	0.96	0.83	0.5	0.47	0.48	0.49	1	0.8	0.6
245	0.44	0.46	0.41	0.77	0.97	0.3	0.43	0.43	0.44	0.8	1	0.4
246	0.44	0.42	0.47	0.62	0.41	0.92	0.46	0.45	0.44	0.58	0.38	1

Table 19. Results of model similarity from Trout dataset

Machine Learning	IBK						J48					
Feature Selection	None			CFS			None			CFS		
Split (%)	100	70	30	100	70	30	100	70	30	100	70	30
Model	121	122	123	124	125	126	271	272	273	274	275	276
121	1	0.84	0.62	0.49	0.49	0.47	0.98	0.84	0.63	0.47	0.48	0.49
122	0.84	1	0.46	0.48	0.5	0.47	0.84	1	0.47	0.46	0.47	0.48
123	0.62	0.46	1	0.47	0.46	0.49	0.6	0.46	0.99	0.5	0.48	0.48
124	0.49	0.48	0.47	1	0.84	0.63	0.48	0.49	0.49	0.97	0.84	0.65
125	0.49	0.5	0.46	0.84	1	0.47	0.49	0.5	0.48	0.81	0.98	0.48
126	0.47	0.47	0.49	0.63	0.47	1	0.46	0.47	0.49	0.64	0.49	0.99
271	0.98	0.84	0.6	0.48	0.49	0.46	1	0.84	0.61	0.45	0.46	0.47
272	0.84	1	0.46	0.49	0.5	0.47	0.84	1	0.48	0.46	0.48	0.48
273	0.63	0.47	0.99	0.49	0.48	0.49	0.61	0.48	1	0.49	0.5	0.49
274	0.47	0.46	0.5	0.97	0.81	0.64	0.45	0.46	0.49	1	0.84	0.63
275	0.48	0.47	0.48	0.84	0.98	0.49	0.46	0.48	0.5	0.84	1	0.49
276	0.49	0.48	0.48	0.65	0.48	0.99	0.47	0.48	0.49	0.63	0.49	1

Clustering of Multiple Microarray Experiments Using Information Integration

Elena Kostadinova[1], Veselka Boeva[1], and Niklas Lavesson[2]

[1] Technical University of Sofia, Branch Plovdiv,
Computer Systems and Technologies Department, 4400 Plovdiv, Bulgaria
{elli,vboeva}@tu-plovdiv.bg
[2] School of Computing, Blekinge Institute of Technology,
SE-371 79 Karlskrona, Sweden
Niklas.Lavesson@bth.se

Abstract. In this article, we study two microarray data integration techniques and describe how they can be applied and validated on a set of independent, but biologically related, microarray data sets in order to derive consistent and relevant clustering results. First, we present a cluster integration approach, which combines the information containing in multiple data sets at the level of expression or similarity matrices, and then applies a clustering algorithm on the combined matrix for subsequent analysis. Second, we propose a technique for the integration of multiple partitioning results. The performance of the proposed cluster integration algorithms is evaluated on time series expression data using two clustering algorithms and three cluster validation measures. We also propose a modified version of the Figure of Merit (FOM) algorithm, which is suitable for estimating the predictive power of clustering algorithms when they are applied to multiple expression data sets. In addition, an improved version of the well-known connectivity measure is introduced to achieve a more objective evaluation of the connectivity performance of clustering algorithms.

Keywords: Microarray Gene Expression Data, Gene Clustering, Data Integration, Cluster Validation.

1 Introduction

In recent years, DNA microarray technology offers the ability to screen the expression levels of thousands of genes in parallel under different experimental conditions or time points [24]. All these measurements contain information about many different aspects of gene regulation and function, ranging from understanding the global cell-cycle control of microorganisms [21, 23], to cancer in humans [1, 8]. Gene clustering is one of the most important microarray analysis tasks when it comes to extracting meaningful information from expression profiles. In general, clustering is the process of grouping data objects into sets of disjoint classes called clusters, so that objects in the same cluster are more similar to each other than objects in the other clusters, given a reasonable measure of similarity. In the context of microarray analysis, clustering

C. Böhm et al. (Eds.): ITBAM 2011, LNCS 6865, pp. 123–137, 2011.

algorithms have been used to divide genes into groups according to the degree of their expression similarity. Such a grouping may suggest that the respective genes are correlated and/or co-regulated, and moreover that the genes could possibly share a common biological role [21, 23]. Therefore, gene expression data clustering arguably has numerous useful applications as it provides an important source of biological inference, for example, to predict the function of presently unknown genes based on the known functions of genes that are located in the same cluster [27, 31].

Presently, with the increasing number and complexity of available gene expression data sets the combination of data from multiple microarray studies addressing a similar biological question is gaining high importance. In general, the integration and evaluation of multiple data sets promise to yield more reliable and robust results since these results are based on a larger number of samples and the effects of individual study-specific biases are weakened. Consequently, single set analysis will be too restrictive to draw a general inference from a particular gene clustering result. Such limitations and inconsistencies of the standard clustering techniques have motivated our research, which is concerned with how to apply clustering algorithms to a set of independent, but biologically related, microarray data sets in order to derive general, consistent and relevant conclusions. Generally, our strategy is based on two integration techniques: *hierarchical merge* [30] and *hybrid integration* [2] that allow us to combine multiple expression data sets in order to obtain one overall (expression or similarity) matrix. The values of this combined matrix can be interpreted as consensus values supported by all the experiments.

In this article, we show that the aforementioned data integration techniques can be applied to both interpretations of the problem of deriving clustering results from a set of gene expression matrices, that is: 1) information containing in different data sets may be combined at the level of expression (or similarity) matrices and then cluster; 2) given multiple clusterings, one per each data set, find a combined clustering. For example, the combined expression (or similarity) matrix can be passed to the corresponding clustering algorithm for subsequent analysis. Another possibility is to use the same combined (expression or similarity) matrix to initialize k cluster centers. Then some partitioning algorithm can be applied to each expression matrix and the obtained partitions derived separately for each involved data set can ultimately be aggregated into one overall partitioning result.

2 Related Work

Different microarray combination techniques and meta-analysis studies have been published in the bioinformatics literature [5, 7, 12, 13, 16]. For instance, [5] and [13] used the meta-analysis approach, where results are combined on an interpretative level, to integrate microarray experiments for the detection of differential gene expression. Other integration technique, considered in [16], presented a model that uses inter-gene information. Gilks *et al.* [7] proposed a data fusion method based on multivariate regression, which aims at producing a fused data matrix.

Two methods for direct integration analysis of gene information containing in multiple microarray experiments have been proposed in [2, 3, 30]. The first method introduced by Tsiporkova & Boeva [30] uses a hierarchical merge procedure in order to fuse together the multiple-experiment time series expression data. The second method applies a hybrid aggregation algorithm [28] to integrate gene relationships (distances) across different experiments and platforms [2, 3]. It produces a single matrix, consisting of one overall distance for each gene pair. The values of this matrix have been demonstrated to capture more realistically the gene correlation structure than the corresponding average values.

The problem of clustering when multiple microarray data sets are involved can be defined in two different ways: 1) information containing in different data sets may be combined at the level of expression (or similarity) matrices and then cluster; 2) given multiple clusterings, one per each data set, find a combined clustering solution. There are several approaches to combine within a clustering process the information contained in different gene representations [11, 27]. One representation considered in this case are gene expression data received from a single microarray experiment. The other representation is Gene Ontology containing knowledge about *e.g.*, gene functions gained throughout the years. Some studies concerning different ways of combining gene information representations at the level of similarity matrices have been proposed [11, 18]. Methods for the combination of the partitions derived for each representation separately have been considered in [25, 26].

In this work, we study two different microarray data integration techniques: hierarchical merge [30] and hybrid integration [2, 3], and describe how they can be applied and validated on a set of gene expression matrices in order to derive consistent and relevant clustering results.

3   Techniques for Clustering of Multiple Microarray Data Sets

We will now present our method by first introducing partitioning in general and then by describing how to perform clustering on multiple microarray data sets.

Three partitioning algorithms are well known and commonly used for microarray analysis to divide the data objects into k disjoint clusters [20]: k-means, k-medians and k-medoids clustering. All these methods start by initializing a set of k cluster centers, where k is preliminarily determined. Then, each object of the data set is assigned to the cluster whose center is the nearest, and the cluster centers are recomputed. This process is repeated until the objects inside every cluster become as close to the center as possible and no further object item reassignments take place. The three partitioning methods in question differ in how the cluster center is defined. In k-means clustering, the cluster center is defined as the mean data vector averaged over all objects in the cluster. Instead of the mean, in k-medians the median is calculated for each dimension in the data vector. Finally, in k-medoids clustering, which is a robust version of the k-means, the cluster center is defined as the object with the smallest sum of distances to the other objects in the cluster, *i.e.*, this is the most centrally located point in a given cluster.

Now, we show how two microarray data integration techniques (hybrid integration and hierarchical merge) can be applied to both definitions of the problem of deriving clustering results from a set of gene expression matrices. We propose a cluster integration approach, which combines the information containing in multiple microarray experiments at the level of expression or distance matrices and then applies a partitioning algorithm on the combined matrix. In addition, an algorithm for the integration of multiple partitioning results is introduced.

First, we consider how a partitioning algorithm can be applied to a set of multiple microarray experiments. Assume that a particular biological phenomenon is monitored in a few high-throughput experiments under n different conditions. Each experiment i ($i = 1, 2,..., n$) is supposed to measure the gene expression levels of m genes in n_i different experimental conditions or time points. Thus a set of n different data matrices $M_1, M_2,..., M_n$ will be produced, one per experiment. Initially, some integration procedure (hybrid integration or hierarchical merge) is applied to transform the set of input matrices $M_1,..., M_n$ into a single matrix, which values can be interpreted as consensus values supported by all the experiments. Then, the overall matrix is passed to the corresponding clustering algorithm for subsequent analysis. In our work, this idea is demonstrated by implementing two partitioning algorithms: k-medoids and k-means. Since the k-medoids clustering algorithm is suitable for cases in which the distance matrix is known but the original data matrix is not available, the hybrid integration procedure [2] is used to combine the quadratic distance (similarity) matrices, generated per each considered data set. On the other hand, k-means clustering method requires an original expression data matrix as input data set and thus, the hierarchical merge algorithm [30] is used to merge the expression profiles from the original input matrices. Finally, the obtained integrated similarity (or fused expression) matrix is passed to k-medoids (or k-means) clustering algorithm for subsequent analysis. Thus, we propose two different cluster integration techniques, which are based on the combination of information containing in the involved data sets at the level of expression and similarity matrices, respectively.

Notice that, by applying information about the quality of the microarrays, weights may be assigned to the experiments and can be further used in the integration process in order to obtain more realistic overall values. A novel data transformation method aiming at multi-purpose data standardization and inspired by gene-centric clustering approaches was proposed in [4]. This standardization algorithm may be used to evaluate the quality of the considered microarrays. For instance, it can be applied to each expression matrix and the number of standardized genes can eventually be considered as a measure of the quality of a microarray. Thus each expression matrix i can be assigned a weight w_i: $w_i = \left(1 - m_i \middle/ \sum_{j=1}^{n} m_j \right) \middle/ (n - 1)$, where n is the number of the integrated experiments and m_i is the number of standardized genes in experiment i.

In this section, we also introduce an algorithm for integrating partitioning results derived from multiple microarray data sets. Initially, the quality of the aforementioned expression matrices can be evaluated by using the weighting scheme discussed above. This step will result in the assignment of a weight w_r to each expression matrix r ($r = 1,..., n$), i.e., all data sets will contribute to the final partitioning result, albeit to different extents. The next step is to initialize the k cluster centers: $C_1,..., C_k$, by using the information contained in the included data sets in an integrated manner. This step can be performed either by applying the hybrid integration procedure or the hierarchical merge procedure on the distance or expression matrices. Whereas the latter is suitable for k-means, the former is better suited for k-mediods. Another possibility altogether, which is suitable for an arbitrary partitioning algorithm, is to use a $m \times (n_1 + n_2 + ... + n_n)$ matrix constructed by concatenating matrices $M_1,..., M_n$.

The selected partitioning algorithm can then be applied to each expression matrix, which will generate a set of $k \times m$ partition matrices: $P_1, \ldots, P_n$. Each partition matrix may be represented as $P_r = \{ p_{ij}^r \}$, where p_{ij}^r is the membership of gene j ($j = 1, \ldots, m$) to the ith ($i = 1, \ldots, k$) cluster in partition matrix r ($r = 1, \ldots, n$). To integrate the different partitioning results into a final partition matrix P, we propose the following aggregation schema. First, the combined matrix $P = \{p_{ij}\}$, $i = 1, \ldots, k$ and $j = 1, \ldots, m$, obtained by summing the partition matrices over all expression data sets, is constructed as follows: $p_{ij} = \sum_{r=1}^{n} p_{ij}^r w_r$, where w_r denotes a weight of matrix r. The proposed method uses all the partition matrices to create the combined result by summing the weighted memberships over all partitioning results. Ultimately, the final partition matrix P is constructed by assigning each gene to the cluster with the highest summed membership. In the case of tie, the gene in question is randomly assigned to one of the tied clusters.

4   Experimental Setup

The performance of proposed cluster integration techniques has been evaluated on time-series microarray data sets using k-means and k-medoids clustering methods and three different cluster validation measures.

4.1   Data

The clustering results of the aforementioned cluster integration algorithms are evaluated on artificial (where true clustering is known) and real (where true clustering is unknown) gene expression time series data obtained from a study examining the global cell-cycle control of gene expression in fission yeast *S. pombe* [23]. The study includes nine different expression matrices. In the pre-processing phase the rows with more than 25% missing entries have been filtered out from each expression matrix and any other missing expression entries have been imputed by the DTWimpute algorithm [29]. In this way nine complete matrices have been obtained. Further, a set of 2935 overlapping genes has been found across all data sets. Subsequently, the time expression profiles of these genes have been extracted from the original data matrices. In that way nine new matrices, which are our real test data sets, have been built.

Rustici *et al.* [23] identified 407 genes as cell-cycle regulated, which are separated into 4 clusters. The genes that are neither assigned to any cluster nor presented in the intersection of the nine original data sets have been removed. The latter produces a subset of 219 genes. Subsequently, the time expression profiles of these genes have been extracted from the original data matrices and thus nine new matrices, which form our artificial data sets with a known clustering solution, have been constructed.

4.2   Cluster Validation Measures

One of the most important issues in cluster analysis is the validation of clustering results. Essentially, the cluster validation techniques are designed to find the partitioning that best fits the underlying data, and should therefore be regarded as a key tool in the interpretation of clustering results. The data mining literature provides a range of different cluster validation measures, which are broadly divided into two major categories: external and internal [14]. External validation measures have the benefit of providing an independent assessment of clustering quality, since they validate a clustering result by comparing it to a given external (gold) standard. However, an external gold standard is rarely available, and for this particular reason, the external validation methods will not be considered here. Internal validation techniques, on the other hand, avoid the need for using such additional knowledge, but have the alternative problem to base their validation on the same information used to derive the clusters themselves. Additionally, some authors consider a third approach of clustering validity, which is based on relative criteria [9]. The basic idea is to evaluate the clustering result by comparing clustering solutions generated by the same algorithm but with different input parameter values. A number of validity indices have been defined and proposed for each of the above approaches. Furthermore, internal measures can be split with respect to the specific clustering property they reflect and assess to find an optimal clustering scheme: *compactness, separation, connectedness*, and *stability* of the cluster partitions. *Compactness* evaluates the cluster homogeneity that is related to the closeness within a given cluster. *Separation* demonstrates the opposite trend by assessing the degree of separation between individual groups. The third type of internal validation measure (*connectedness*) quantifies to what extent the nearest neighboring data items are placed into the same cluster. The *stability* measures evaluate the consistency of a given clustering partition by clustering from all but one experimental condition. The remaining condition is subsequently used to assess the predictive power of the resulting clusters by measuring the within-cluster similarity in removed experiment. A detailed summary of different types of validation measures can be found in [10]. Since the aforementioned criteria are inherently related in the context of both classification and clustering problems, some approaches have been presented which try to evaluate multiple of these criteria or that try to improve one criterion by optimizing another. For example, Lavesson & Davidsson [19] empirically analyzed a practical multi-criteria measure based on estimating the accuracy, similarity, and complexity properties of classification models. In the context of the presented study, accuracy referred to the predictive power of the model, similarity referred to the extent to which instances of the same type are situated in proximity of each other, and complexity referred to the number of partitions or clusters in the model. The similarity property was divided into two components; if an instance of a certain type was detected outside the associated cluster, the distance between the instance and the cluster border contributed negatively to the measure. However, if an instance of a certain type was situated within the associated cluster, the distance to the cluster border contributed positively to the measure. Additionally, Davidsson [6] introduced an approach for controlling the generalization during model generation on (training) data to ensure that instances of unknown class/type would not be classified as belonging to known (learned) clusters or classes. To understand the importance of this, consider the situation in which a certain data set has been used to generate two clusters and that a validation data set is then analyzed to estimate the performance of the clustering solution. Moreover, assume that the first data set features instances of two different types while the second data set features a third type of instance in addition to the two known types. Typically, many clustering and classification solutions would assign instances of the third type to one of the generated clusters even though they represent other instance types. The approach presented by Davidsson introduces a confidence area for each cluster, which encompasses the known instances of the cluster, *e.g.*, the instances used to generate the cluster, as well as an additional area outside of the cluster, for which the size is determined by a confidence factor. The more distant an instance is from a cluster center, the more likely it is regarded as not belonging to the cluster (the probability of it belonging to the cluster decreases as the distance increases). Thus, given a certain confidence factor, *e.g.*, $\alpha = 0.05$, an instance may be regarded as belonging to specific cluster even if it is outside of the cluster since it is inside the outer area. However, if the instance was outside the outer area, *e.g.*, if the confidence factor was higher, the instance would be categorized as being of unrecognized type, *i.e.*, not belonging to any of the generated clusters.

Since none of the clustering algorithms perform uniformly best under all scenarios, it is not reliable to use a single validation measure, but instead to use a few that reflect various aspects of a partitioning. In this sense, we have implemented three different validation measures for estimating the quality of clusters, produced by integration analysis of multiple microarray matrices. Based on the above mentioned classification, we have selected one validation measure from each category: for assessing compactness and separation properties of a partitioning - *Silhouette Index* (*SI*); for assessing connectedness - *Connectivity*; and the *Figure of Merit* (*FOM*) has been used for predictive power assessment.

The subsequent subsections not only demonstrate the technical details of the selected validation methods, but also introduce some novel concepts of their application in the case of integrated data analysis. Initially, the validation algorithm *FOM* has been modified in order to estimate the predictive power of clustering algorithms that are applied to multiple expression data sets. Subsequently, an improved revision of the well-known *Connectivity* measure is proposed for more objective evaluation of the connectivity performance of clustering algorithms.

4.2.1 Figure of Merit

Yeung *et al.* [32] defined *figures of merit*, which are estimators of the predictive power of clustering algorithms, to assess the quality of clustering results. The predictive power is estimated by removing one experiment from the data matrix, clustering genes based on the remaining data, and measuring the within-similarity of expression values in the removed experiment. The figure of merit is the root mean square deviation in the left-out condition of the individual gene expression levels relative to their cluster mean.

This approach is analogous to *leave-one-out cross-validation*, which is a common predictive performance estimation method in statistics and machine learning. In leave-one-out cross-validation, the objective is to estimate the predictive accuracy of a *classifier*, which is a function or mapping from a set of data instances to a set of classes. Commonly, *supervised learning* algorithms are used to generate classifiers [17]. For this machine learning task, the main idea is to apply a learning algorithm to detect patterns in a data set (inputs) that are associated with known class labels (outputs) in order to automatically create a generalization; that is, a classifier. Under the assumption that the known data properly represent the complete problem studied, it is further assumed that the generated classifier will be able to predict the classes of novel data instances. This is in contrast to the approach in *FOM*, where no prior information of the genes is assumed when clustering or when evaluating the quality of the clustering results. Instead, *FOM* (which can arguably be regarded as an estimator of the predictive power of a clustering algorithm) is defined to assess the quality of the clustering results.

We show herein how this approach can be used for validation of clustering results, which are obtained from integrated microarray data. Consider a set of n different expression matrices $M_1,\ldots, M_n$ and each matrix i ($i = 1,\ldots, n$) is supposed to measure the gene expression levels of m genes in n_i different experimental conditions or time points. Suppose the data from matrices $M_1,\ldots, M_{(r-1)}, M_{(r+1)}, \ldots, M_n$ are combined by applying some integration (fusion) technique. Further a clustering algorithm is applied to the integrated data and matrix M_r is used to estimate the predictive power of the algorithm. Assume k clusters, $C_1,\ldots, C_k$, are obtained. Let g_{jt} be the expression level of gene j under condition (time point) t in matrix M_r and $\mu_{C_{it}}$ be the average expression level in condition t of genes in cluster C_i. Then the *figure of merit*, FOM (M_r, k), can be defined as the root mean square deviation in the left-out matrix M_r of the individual gene expression profiles relative to their cluster means:

$$FOM(M_r,k) = \sqrt{\frac{1}{m} \sum_{i=1}^{k} \sum_{j \in C_i} \sum_{t=1}^{n_r} (g_{jt} - \mu_{C_{it}})^2}. \qquad (1)$$

Each of n matrices can be used as the left-out expression matrix. The *aggregate figure of merit*, $FOM(k) = \sum_{r=1}^{n} FOM(M_r, k)$, is an estimate of the total predictive power of the algorithm over all the matrices for k clusters. Yeung *et al.* [32] also defined the *adjusted figure of merit*, $FOM(k)/\sqrt{(m-k)/m}$, which corrects the tendency of decreasing the *FOM* due to increase the number of clusters.

4.2.2 Connectivity

Connectivity captures the degree to which genes are connected within a cluster by keeping track of whether the neighbouring genes are put into the same cluster [10]. Define $m_{i(j)}$ as the jth nearest neighbour of gene i, and $x_{im_{i(j)}}$ be zero if i and j are in the same cluster and $1/j$ otherwise. Then for a particular clustering solution $C_1,\ldots, C_k$ of matrix M_r, which contains the expression values of m genes (rows) in n_r different experimental conditions or time points (columns), the *connectivity* is define as $Conn(M_r,k) = \sum_{i=1}^{m} \sum_{j=1}^{n_r} x_{im_{i(j)}}$.

The neighbourhood size is a required parameter for this validation algorithm and usually it is preliminarily fixed. For example, in the above equation it equals to the number of columns (n_r). The fact that the number of neighbouring genes is preliminarily defined and does not depend on the degree of similarity between the expression profiles, this may lead to a rather stochastic selection of neighbouring genes for validation. For instance, it may happen that the expression profiles of some neighbouring genes are rather distant from the profile of a certain gene since these profiles have only been considered to reach the required fixed number of genes.

In view of the above we propose herein a validation process, which uses for each gene a varying number (based on the degree of similarity) of neighboring gene profiles. For this purpose we have developed a dedicated algorithm. The latter is employed for generating a list of genes with expression profiles that exhibit at least minimum relative (preliminarily defined) similarity in terms of some distance measure to the expression profile of the currently considered gene.

The connectivity has a value between zero and infinity and should be minimized.

4.2.3 Silhouette Index

Silhouette Index [22] is a cluster validity index that is used to judge the quality of any clustering solution (partition) C_1, C_2, ..., C_k. Suppose a_i represents the average distance of gene i from the other genes of the cluster to which the gene is assigned, and b_i represents the minimum of the average distances of gene i from genes of the other clusters. Then the *Silhouette Index* (*SI*) of matrix M_r, which contains the expression profiles of m genes, is defined as: $s(M_r,k)=(1/m)\cdot\sum_{i=1}^{m}(b_i-a_i)/\max\{a_i,b_i\}$.

The values of Silhouette Index vary from -1 to 1 and higher value indicates better clustering results. It reflects the compactness and separation of clusters.

4.3 Estimation of the Number of Clusters

A validation approach, which is based on relative criteria and does not involve statistical tests, was discussed in [9]. Its general idea is to choose the best clustering scheme of a set of defined schemes according to a pre-specified criterion.

The partitioning algorithms contain the number of clusters (k) as a parameter and their major drawback is the lack of prior knowledge for that number to construct. Unfortunately, determining a correct, or even suitable, k is a difficult, if not impossible, problem in a real microarray data set. This is especially true when there is an absence of a well-grounded statistical model [15]. For such cases, researchers usually try to generate clustering results for a range of different numbers of clusters and subsequently assess the quality of the obtained clustering solutions. For example, some of the internal validation measures that we have implemented in the presented study (see Section 4.2) can be used as validity indices to identify the best clustering scheme. Thus, we have run the two clustering algorithms for all values of k between 2 and 20 and plot the values of the selected indices (*FOM* and *SI*) obtained by each k as the function of k (see Supplementary Material[1] Fig. 1 and Fig. 2 (a)). Notice that both indices decrease as the number of clusters increase. Therefore we search for the values of k at which a significant local change in value of the index occurs [9]. These values are different for the considered validity indices and clustering algorithms, and vary between 5 and 15. That is why a different number of clusters in this range has been selected and then used to evaluate the performance of both clustering algorithms.

5   Results and Discussion

In this section, we compare the performance of k-means and k-medoids clustering algorithms on the individual and integrated matrices of the data sets described in Section 4.1 by using three cluster validation measures: *FOM*, *SI* and *Connectivity*.

---

[1] Supplementary data are available at: http://cst.tu-plovdiv.bg/bi/Supplementary_Material.pdf

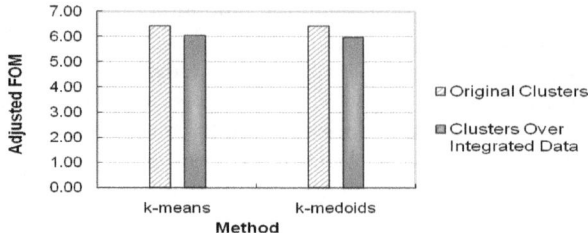

Fig. 1. *Adjusted FOM* results generated by the modified *FOM* algorithm on Rustici *et al.* artificial data sets

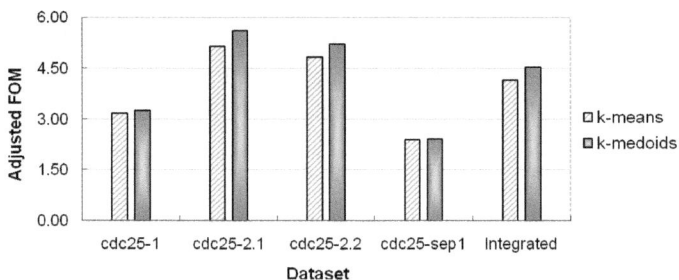

Fig. 2. *Adjusted FOM* values on individual and integrated *cdc* matrices from Rustici *et al.* data

Fig. 1 depicts the *Adjusted FOM* values generated from the known clustering solution versus those obtained by applying the clustering algorithms on the integrated data. The experiment is performed using the artificial data sets from Rustici *et al.* [23]. The results are calculated by the proposed modified *FOM* algorithm (see Eq. 1), which is especially well suited for the validation of integrated data. It can be seen that the original clusters have higher *Adjusted FOM* values (respectively, worse performance) than those generated on the clusters that are obtained from the combined data.

Fig. 2 compares the *Adjusted FOM* results generated on the individual and integrated distance (expression, respectively) *cdc* data matrices using $k = 10$ clusters. Note that the results for the individual data sets in Fig.2 are obtained by applying the standard *FOM* approach [32], while the values for the integrated *cdc* data are produced by using the modified *FOM* algorithm (explained in Section 4.2.1). As one can notice the scores obtained on the integrated data sets are trade-off between those given on the individual matrices. The latter implies that the specific characteristics of the individual microarray experiments are equally reflected on the integrated data.

Fig. 3 presents the *SI* values generated on each individual Rustici data set and on the combined expression and distance matrices. In addition, the *SI* values that are produced on the matrix obtained by averaging the corresponding values of the individual distance matrices are depicted in Fig. 3 (right). The clustering methods have been studied for two different cluster numbers $k = 7, 10$. It can be seen in Fig. 3 (left) that the *SI* scores produced on the fused expression matrix are in most cases (especially, for $k = 7$) better than those obtained on the individual ones. Moreover, they are comparable to those given on the best performing individual data set (elu-cdc10). In contrast, the results obtained on the individual data sets outperform the results generated on the integrated distance matrix. The latter may be due to the fact that the values in the integrated distance matrix are calculated by applying the hybrid aggregation algorithm, which produces trade-off values agreed between the individual distance matrices. These compromised distance values certainly affect the performance of k-medoids algorithm with respect to cluster separation. However, the integrated distance matrix performs better than the corresponding averaged matrix for both values of k.

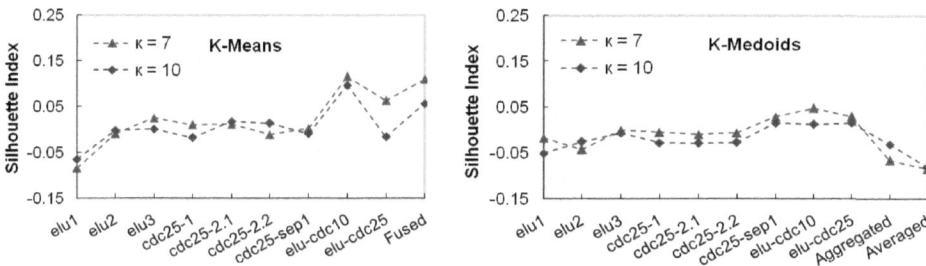

Fig. 3. *SI* values obtained on the individual, fused expression, integrated and averaged distance Rustici *et al.* data

An improved version of connectivity measure, which uses for each gene a varying number (based on the degree of similarity) of neighbouring gene profiles, is discussed in Section 4.2.2. Thus the results in Supplementary Fig. 3 (a) are obtained by applying the standard connectivity approach [10], while those in Supplementary Fig. 3 (b) are produced by the proposed modified version. The standard connectivity approach has been investigated on the integrated distance and fused expression matrices of all the test data sets for different numbers of neighbouring genes. It can be seen that the lowest scores (connectivity ≤ 3000) are generated for neighbours in the range [1, 30] for k-means and [1, 4] for k-medoids, respectively. Supplementary Fig. 3 (b) shows the connectivity values on the same combined matrices for a range of different similarity cut-offs, which are determined for each gene as a percentage of the average distance to its expression profile. The best results (connectivity = 0) are produced for similarity cut-off up to 25% of the average distance for k-medoids and up to 15% for k-means, respectively. We can notice in Supplementary Fig. 3 (a) that the connectivity performance of both clustering methods is similarly influenced by the number of neighboring genes considered as a parameter of the partitioning algorithms, while k-means and k-medoids perform differently with regard to the neighbourhood size in Supplementary Fig. 3 (b). The latter supports our belief that the modified connectivity measure proposes a more realistic assessment of the connectivity performance of clustering algorithms than the standard approach.

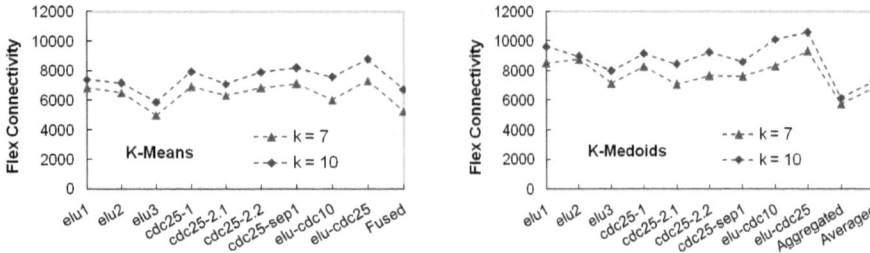

Fig. 4. *Connectivity* values calculated on the individual and integrated Rustici *et al.* data sets using a varying number of neighboring profiles

Fig. 4 benchmarks the calculated connectivity scores generated on each individual Rustici *et al.* data set and on the fused expression (integrated distance, respectively) matrix using a varying number (based on the degree of similarity) of neighboring gene profiles. The results have been calculated for similarity cut-off determined as 50% of the average distance value. Results for another similarity cut-off can be found in Supplementary Material (see Supplementary Fig. 4). In addition, connectivity scores obtained on the same data sets by the standard connectivity approach [10] are presented in Supplementary Fig. 5. The latter figure demonstrates that the values generated on the fused expression matrix outperforms those obtained on the individual ones while the results on the aggregated distance matrix are compatible to those of the best performing individual experiment. However, as it can be seen in Fig. 4, the connectivity values produced by the modified approach on both combined (expression and distance) matrices outperform those given on the individual data sets. Additionally, the connectivity results generated by the integrated distance matrix are again better than the values obtained by averaging the corresponding values of the individual distance matrices.

We have used the weighting scheme considered in Section 3 to evaluate the quality of the microarray matrices in our test corpus. This results in the assignment of a weight to each expression matrix. Then the weighted version of the hybrid aggregation algorithm has been applied to the individual distance matrices in order to generate a single integrated matrix. Fig. 6 in Supplementary Material benchmarks the *SI* score generated on this matrix and the corresponding values on the individual, integrated and averaged matrices. Although, the weighted version of the hybrid integration algorithm is observed to have the best performance in terms of *SI* index it does not substantially improve the performance of *k*-medoids clustering algorithm. The latter may be explained by the fact that the used test data sets are not significantly different in their quality. Perhaps, other microarray data should be involved in the further study of the proposed weighting scheme. Different weighting schemes may be considered, as well.

The technique for integrating multiple partitioning results that was proposed in Section 3 has been studied by applying a partitioning algorithm (*k*-means and *k*-medoids) to each microarray data set from Rustici *et al.* and then aggregating the partitions derived separately for each input matrix into a single partitioning result. The performance of this partition integration technique has been evaluated using two cluster validation measures: *Silhouette Index* and *Connectivity*. Fig. 5 benchmarks the *SI* scores generated by independent clustering of the individual data sets against the mean across the *SI* values calculated on the individual partitioning results[2], referred to Individual, and the mean of the corresponding *SI* values given on the integrated partition, referred to Integrated. One can observe that the scores obtained by the integrated partition are fairly exceeded by the *SI* values produced on the individual matrices. The identical results presented in Supplementary Material (see Supplementary Fig. 7) are observed for the connectivity index. These results are mainly due to the high percentage of tied genes, *i.e.* those which are randomly assigned in the case of tie (see Section 3). For example, we have found almost 32% tied genes for *k*-means and 34% for *k*-medoids, respectively, using $k = 7$ clusters. The percentages of ties calculated for the both partitioning algorithms by using different cluster numbers are available in Table 1 of Supplementary Material. It can easily be noticed in Fig. 5 that the mean of the *SI* values produced by the individual partitions (Individual) is, in fact, an average of the *SI* scores obtained by independent clustering of individual data sets. However, the clustering performance is surprisingly degraded in the case of aggregating the individual partitions (Integrated), which is clearly caused by the high percentage of random gene assignments.

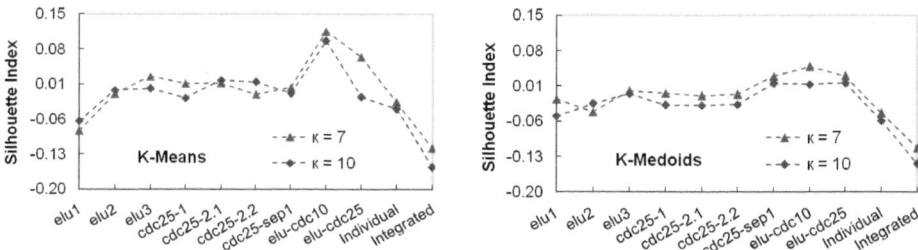

Fig. 5. *SI* scores generated on the individual Rustici *et al.* data sets versus means of *SI* values obtained on individual and integrated partitions, respectively

It is evident that the cluster integration approach, which applies a partition algorithm on the combined (expression or distance) matrix, performs better than those integrating multiple partitioning results, because the former approach always succeeds in finding an agreement between the experiments for a cluster of the gene in question.

6   Conclusion

We have studied two microarray data integration techniques and have shown how they can be applied to both interpretations of the problem of deriving clustering results from a set of gene expression matrices. Initially, we have considered a cluster integration approach, which combines the information containing in multiple

---

[2] These partitions are obtained by initializing the cluster centers, which are common for all the individual matrices.

microarray experiments at the level of expression or distance matrices and then applies a partitioning algorithm on the combined matrix. Furthermore, a technique for the integration of partitioning results derived from multiple microarray data sets has been introduced. The proposed integration techniques have been validated on time series expression data by using two partitioning algorithms and three different cluster validation indices. It has been shown that the application of a partition algorithm on the integrated data yields better performance compared to the approach of aggregating multiple partitions, since the former approach manages to avoid the problem with ties.

Our future plans include the further improvement of the approach of integrating multiple partitioning results by applying other more elaborated methods for resolving the ties. Different techniques for the initialization of cluster centers can also be studied. In addition, the proposed approaches can be investigated in other types of microarray data and platforms. After more elaborate evaluation and validation of the methodology proposed in this article, the aim is to develop an analysis tool in which it would be possible to integrate microarray data sets, obtain clustering results, and then get visual feedback on possible functions of previously unknown genes.

Acknowledgements. The authors would like to thank ASIC Depot Ltd. (http://www.asicdepot.com), Bulgaria for the financial support which made possible the participation in this conference.

References

1. Alizadeh, A., et al.: Distinct types of diffuse large B-cell lymphoma identified by gene expression profiling. Nature 403, 503–511 (2000)
2. Boeva, V., Kostadinova, E.: A Hybrid DTW based method for integration analysis of time series data. In: ICAIS 2009, Austria, pp. 49–54 (2009)
3. Boeva, V., Kostadinova, E.: An Adaptive Approach for Integration Analysis of Multiple Gene Expression Datasets. In: Dicheva, D., Dochev, D. (eds.) AIMSA 2010. LNCS, vol. 6304, pp. 221–230. Springer, Heidelberg (2010)
4. Boeva, V., Tsiporkova, E.: A Multi-purpose Time Series Data Standardization Method. In: Sgurev, V., Hadjiski, M., Kacprzyk, J. (eds.) Intelligent Systems: From Theory to Practice. SCI, vol. 299, pp. 445–460. Springer, Heidelberg (2010)
5. Choi, J.K., et al.: Combining multiple microarray studies and modeling interstudy variation. Bioinformatics 19, i84–i90 (2003)
6. Davidsson, P.: Coin Classification Using a Novel Technique for Learning Characteristic Decision Trees by Controlling the Degree of Generalization. In: Ninth International Conference on Industrial & Engineering Applications of Artificial Intelligence & Expert Systems, pp. 403–412. Gordon and Breach Science Publishers, New York (1996)
7. Gilks, W.R., Tom, B.D.M., Brazma, A.: Fusing microarray experiments with multivariate regression. Bioinformatics 21(2), ii137–ii143 (2005)
8. Golub, T., et al.: Molecular classification of cancer: class discovery and class prediction by gene expression monitoring. Science 286, 531–537 (1999)
9. Halkidi, M., Batistakis, Y., Vazirgiannis, M.: On clustering validation techniques. Journal of Intelligent Information Systems 172(3), 107–145 (2001)
10. Handl, J., et al.: Computational cluster validation in post-genomic data analysis. Bioinformatics 21, 3201–3212 (2005)

11. Havens, T.C., et al.: Fuzzy cluster analysis of bioinformatics data composed of microarray expression data and Gene Ontology annotations. In: North American Fuzzy Information Processing Society, pp. 1–6 (2008)
12. Hermans, F., Tsiporkova, E.: Merging microarray cell synchronization experiments through curve alignment. Bioinformatics 23, e64–e70 (2007)
13. Hu, P., et al.: Integrative analysis of multiple gene expression profiles with quality-adjusted effect size models. BMC Bioinformatics 6, 128 (2005)
14. Jain, A.K., Dubes, R.C.: Algorithms for clustering data. Prentice Hall, Englewood Cliffs (1988)
15. Jain, A.K., Moreau, J.V.: Bootstrap technique in cluster analysis. Pattern Recognit. 20, 547–568 (1987)
16. Kang, J., Yang, J., Xu, W., Chopra, P.: Integrating heterogeneous microarray data sources using correlation signatures. In: Ludäscher, B., Raschid, L. (eds.) DILS 2005. LNCS (LNBI), vol. 3615, pp. 105–120. Springer, Heidelberg (2005)
17. Kohavi, R.: A study of cross-validation and bootstrap for accuracy estimation and model selection. In: IJCAI (1995)
18. Kustra, R., Zagdanski, A.: Incorporating Gene Ontology in Clustering Gene Expression Data. In: 19th IEEE Symposium on Computer-Based Medical Systems, pp. 555–563 (2006)
19. Lavesson, N., Davidsson, P.: A Multi-dimensional Measure Function for Classifier Performance. In: 2nd IEEE Internat. Conf. on Intelligent Systems, pp. 508–513. IEEE Press, Los Alamitos (2004)
20. MacQueen, J.B.: Some methods for classification and analysis of multivariate observations. In: Proc. Fifth Berkeley Symp. Math. Stat. Prob., vol. 1, pp. 281–297 (1967)
21. Oliva, A., et al.: The cell cycle-regulated genes of Schizosaccharomyces pombe. PLOS 3(7), 1239–1260 (2005)
22. Rousseeuw, P.: Silhouettes: a graphical aid to the interpretation and validation of cluster analysis. Journal of Computational Applied Mathematics 20, 53–65 (1987)
23. Rustici, G., et al.: Periodic gene expression program of the fission yeast cell cycle. Nat. Genetics 36, 809–817 (2004)
24. Schena, M., et al.: Quantitative monitoring of gene expression patterns with a complementary DNA microarray. Science 270, 467–470 (1995)
25. Strehl, A., Ghosh, J.: Cluster Ensembles – A Knowledge Reuse Framework for Combining Multiple Partitions. Journal of Mach. Learning Research 3, 583–617 (2002)
26. Topchy, A., Jain, K., Punch, W.: Clustering ensembles: models of consensus and weak partitions. IEEE Trans. Pattern Anal. Machine Intelligence 27, 1866–1881 (2005)
27. Troyanskaya, et al.: A Bayesian framework for combining heterogeneous data sources for gene function prediction (In S. cerevisiae). Genetics. PNAS 100, 8348–8353 (2003)
28. Tsiporkova, E., Boeva, V.: Nonparametric Recursive Aggregation Process. Kybernetika. J. of the Czech Society for Cybernetics and Inf. Sciences 40(1), 51–70 (2004)
29. Tsiporkova, E., Boeva, V.: Two-pass imputation algorithm for missing value estimation in gene expression time series. JBCB 5(5), 1005–1022 (2007)
30. Tsiporkova, E., Boeva, V.: Fusing Time Series Expression Data through Hybrid Aggregation and Hierarchical Merge. Bioinformatics 24(16), i63–i69 (2008)
31. Xiao, G., Pan, W.: Gene function prediction by a combined analysis of gene expression data and protein–protein interaction data. JBCB 3, 1371–1389 (2005)
32. Yeung, K.Y., Haynor, D.R., Ruzzo, W.L.: Validating clustering for gene expression data. Bioinformatics 17(4), 309–318 (2001)

A High Performing Tool for Residue Solvent Accessibility Prediction

Lorenzo Palmieri[1], Maria Federico[1], Mauro Leoncini[1,2], and Manuela Montangero[1,2]

[1] Dipartimento di Ingegneria dell'Informazione, Università di Modena e Reggio Emilia, Italy
[2] CNR, Istituto di Informatica e Telematica, Pisa, Italy
{lorenzo.palmieri,maria.federico,mauro.leoncini,
manuela.montangero}@unimore.it

Abstract. Many efforts were spent in the last years in bridging the gap between the huge number of sequenced proteins and the relatively few solved structures. Relative Solvent Accessibility (RSA) prediction of residues in protein complexes is a key step towards secondary structure and protein-protein interaction sites prediction. With very different approaches, a number of software tools for RSA prediction have been produced throughout the last twenty years. Here, we present a binary classifier which implements a new method mainly based on sequence homology and implemented by means of look-up tables. The tool exploits residue similarity in solvent exposure pattern of neighboring context in similar protein chains, using BLAST search and DSSP structure. A two-state classification with 89.5% accuracy and 0.79 correlation coefficient against the real data is achieved on a widely used dataset.

1 Introduction

Protein folding is the physical process by which a polypeptide chain folds into its characteristic and functional native structure from a simple sequence of amino acids. It's widely believed that this structure is determined as a whole by the residues sequence. Understanding the key mechanisms of protein folding is for the time being one of the major concern in molecular biology and drug design. However, assessment of solvent accessibility is strongly connected to folding because of the high correlation between hydrophobic forces driving core residues towards a buried exposure state, hence determining the folded structure. Also, solvent accessibility is a strong discriminant for residues lying on the surface, thus becoming likely candidates for being Protein-Protein Interaction sites [1].

As the gap between the number of sequenced proteins and three-dimensional solved structures keeps increasing, many investigation efforts are being made to develop methods able to determine solvent accessibility using only primary sequence data [2,3]. Several exposure state interpretations of residue surface area have been proposed. "Classifiers" ideally divide side-chains exposition area of amino acids in a number of discrete intervals, typically two, three or ten. On the other hand, "real value" approaches describe exposure state by a "continuous" range of values in the [0,1] interval, depending on the exposed surface of the residue. The exposure area is usually computed, for each amino acid, as percent of the maximum area of its side chain that can be exposed to

C. Böhm et al. (Eds.): ITBAM 2011, LNCS 6865, pp. 138–152, 2011.
© Springer-Verlag Berlin Heidelberg 2011

the solvent. Several different threshold values have been used by discrete classifiers described in the literature, with 5%, 10%, 20%, and 25% being the most popular. Depending on this percent value, amino acids are then classified as either buried or exposed on a binary basis, or with discrete exposure levels in case of multiple threshold systems [4].

Several different approaches have been proposed to cope with the solvent accessibility problem: Information Theory [5,6], Bayesian Statistics [7], Probability Profiles [8], Neural Networks [4,9,11,12,13,14,15], Linear Regression [16,17], Support Vector Machines [18,19], Support Vector Regression [20], Look-up Tables [21], meta-methods [22] and many others [23]. However, exploiting sequence similarity to known structures, namely sequence homology, proved to be a substantial improvement strategy for all these methods, both for secondary structure and Solvent Accessibility prediction [9,24]. In many cases sequence homology dramatically improved accuracy of prediction [7,25]. The improvement rate given by this approach is getting more and more tangible with time, by virtue of the thousands new structures solved every year and deposited in the PDB [26].

We developed a software tool for predicting Solvent Accessibility starting from the amino acidic sequence, which exploits the sequence homology information in an efficient and effective way. The underlying algorithm is based on dictionary-like data structures, and takes advantage of information stored in online databases, providing a very high performance on different kinds of datasets, matching the most popular released software tools, and often outperforming them.

2   Tool Overview

Our Relative Solvent Accessibility (RSA) prediction tool is a binary classifier which assigns a *buried* or *exposed* state to each residue of the query sequence. The tool works in two phases, as outlined in Figure 1. Given a query sequence, in the first phase the tool: (a) performs a BLAST homology search in order to obtain a list of sequences homologous to the query, rated by similarity [27]; (b) selects a subset of the returned sequences and fetches the corresponding structure information from the Dictionary of Protein Secondary Structure (DSSP) data bank [28]; (c) computes RSA values for residues using these information and appropriately stores them in pattern-based look-up tables. In the second phase the tool makes the predictions by repeatedly accessing the look-up tables for each residue in the query sequence.

The tool is written in Java and shell scripts, runs under Unix/Linux operating systems, and makes use of Protein-Protein BLAST (v2.2.23+).

In the next sections we will describe each phase in more details.

2.1   Fill-up Phase

Homology Search. The query sequence is aligned, using the local alignment algorithm BLAST, against the PDB Data Bank to obtain a list of the most similar sequences whose structures have already been solved. This list is parsed by PDB-Id and the first N solved structures are fetched from the DSSP Data Bank, where N is a tool parameter (in Section 3.1 we will discuss how to choose a proper value for such parameter). In the following, this set of sequences will be addressed as the set of *hit sequences*.

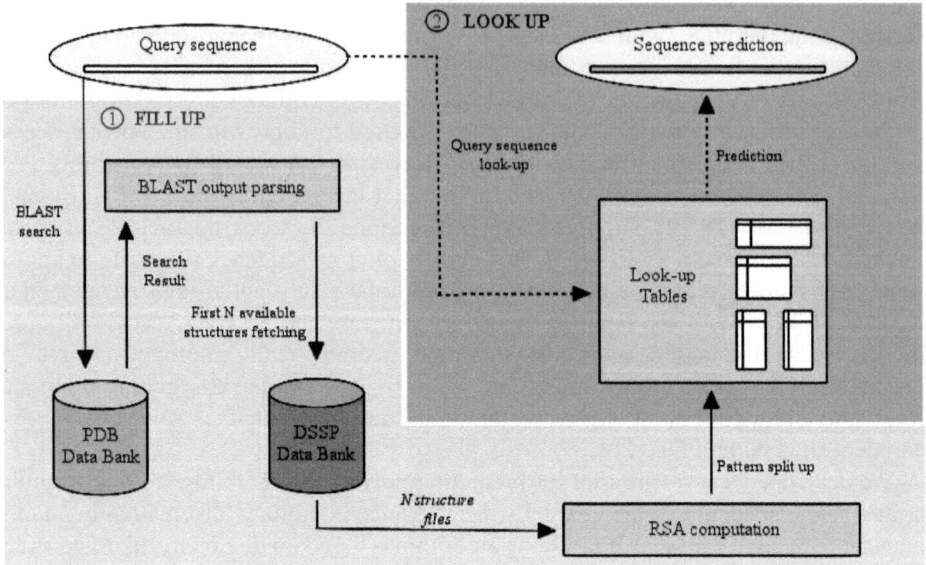

Fig. 1. Two stages prediction workflow. The Fill-up phase includes a BLAST search on PDB known structures, retrieval of structures from DSSP, RSA computation, and Look-up tables creation. The Prediction phase includes looking up the query sequence on the tables, and, finally, residue-by-residue solvent accessibility prediction.

Look-up Tables Creation. The DSSP structure files are parsed to obtain Accessible Surface Area (ASA) values of residues in the hit sequences, then these values are used (details in section 3.1) to obtain residue Relative Solvent Accessibility (RSA) values and used to fill-up some specific look-up tables.

For one such table, the entries correspond to k-tuples of residues. We refer to the residue in the middle position of the tuple as the *central* one, and to the other residues as the *context*. The value stored in a given entry is precisely the average RSA value, computed over all the hit sequences, of its central residue in that context. More formally, an entry could be identified by the pair (central,context) = $(r, \langle \alpha, \beta \rangle)$ (where α and β are oligopeptides of total length $k-1$), and the value stored therein as the average of the RSA values of residue r computed over all k-tuples $\alpha r \beta$ appearing in the hit sequence set.

In particular, the tool creates the following four tables:

- 2P2N, standing for "2 Previous 2 Next", is a 21×21^4 table with the 20 standard amino acids on rows (plus a generic X amino acid sometimes found in DSSP structures) and 21^4 columns, representing all the possible four residues *context* surrounding the central residue (two residues before, two after, corresponding to the oligos α and β in the notation used above); each entry contains the average RSA value for the central amino acid when surrounded (in the hit sequences) by the specific context represented by the column index.

- 1P1N, standing for "1 Previous 1 Next", is a 21×21^2 table, that stores, analogously to the previous one, the average RSA value when the context consists of only two residues (one before, one after).
- 1P and 1N, standing for "1 Previous" and "1 Next", respectively, are two 21×21 tables. Here the context is composed of only one residue, that can be placed before or after the "central" one.

Explorative experiments (data not given here) showed that larger contexts do not significantly improve the tool performance. This result makes sense if we think that the further we move from one residue, the less probable it is that a residue influences the state of the one under consideration (in this paper, we do not take into consideration the possibility that an independent portion of the sequence, at a large and unpredictable distance, might influence the state of the residue because of the 3D structure of the protein).

Although the amount of space required to store our largest table explicitly would not be a problem for modern PCs (around 30MB, using double precision arithmetic), since the tables are typically sparse we chose an implementation based on hashing.

We observe that a similar approach, based on look-up tables, was used in a previous works by Wang et al. and Carugo [21,5,30]. The crucial differences with our work is that tables there were filled up using information derived from the dataset under study, and not from an independent set of homologous sequences. In particular, we look for sequences showing a high degree of similarity with the one to be predicted, under the hypothesis that sequence similarity implies similarities in protein functions and, hence, also structure similarity. Thus, some of the major differences in the design of our tool are: the introduction of homology search for each sequence, decisions on how to use information coming from homology search has to be taken, look-up tables are computed once for each sequence and not once for the entire dataset.

2.2   Look-Up Phase

In the prediction process our tool scans the query sequence residue by residue and, for each residue, accesses the look-up tables in a "hierarchical" fashion, starting from 2P2N down to 1P1N until possibly 1P and 1N.

In details, given a specific query residue, the tool uses its four residue context in the query sequence to access the 2P2N table. In case of a hit (*i.e.*, the value associated to that table entry is non-zero), the RSA value stored in the table is assigned to the analyzed residue of the query sequence as predicted RSA value, and the prediction process moves to the next residue. Otherwise, the two residue context is considered and 1P1N table is examined. In case of another miss (*i.e.*, the value in the appropriate 1P1N entry is zero) a one-residue long context is taken into consideration. We arbitrarily decided to access the 1P table first, and in the case of a miss, the 1N table (by further experiments - data not shown - this assumption turned out not to appreciably influence the prediction performance).

After the look-up phase is completed, the tool assigns a state to the residue that might be *buried* or *exposed*. The decision is made according to the so called *exposure threshold* (given as input parameter) on the RSA value associated to query residues: if the RSA value is under the threshold, the residue is classified as buried, otherwise it is labeled as exposed.

In the rare cases of four misses (i.e., a miss in each of the four tables), the exposed or buried state is assigned to the query residue by means of a default value obtained by a Principal Component Analysis (PCA) study of amino acids physiochemical properties [31]. This study suggests to predict standard amino acids exposure state in the following way: buried for A (Ala), C (Cys), F (Phe), I (Ile), L (Leu), M (Met), V (Val), Y (Tyr), W (Trp) and exposed for the others.

3   Experiments

To evaluate our tool, we worked on already solved protein structures. To avoid over-fitting and allow fair comparisons with other tools, the experiments were carried using a minor variant of the algorithm described in the previous section: given the result of the homology search, we discard sequences that show an exact match with the query sequence PDB identifier; *i.e.*, we discard chains strictly related with the query sequence (sometimes the query sequence itself). Note, however, that the number of selected hit sequences is always equal to the parameter N.

3.1   Datasets and Experimental Setup

We ran experiments using the two datasets described below, which are among those most studied.

Dataset 1 (NM215). This dataset consists of 215 non-homologous protein chains (50878 residues) with no more than 25% pairwise-sequence identity and crystallographic resolution < 2.5Å [6].

Dataset 2 (RS126). This dataset contains 126 non-homologous protein chains (23360 residues), again with no more than 25% pairwise-sequence identity [10].

Prediction Evaluation Indicators.   To evaluate the performance of our tool, we used two performance indicators: *accuracy* and *correlation*.

At sequence level, accuracy is simply the percentage of correctly predicted residues over the total number of residues in the sequence. At data set level, accuracy is the average sequence accuracy of the sequences in the dataset. At residue level, accuracy is meant as the percentage of correctly predicted residue occurrences in the dataset, over the total number of occurrences of that particular residue in the dataset.

Correlation is computed by means of Pearson's Correlation Coefficient (PCC). Given an R-residue long protein chain, let o_i and p_i denote the observed o_i and predicted p_i solvent exposure states of residue i, for $i = 1, \ldots, R$. Then the correlation c of the chain is given by:

$$c = \frac{R \sum_i o_i p_i - \sum_i o_i \sum_i p_i}{\sqrt{R \sum_i o_i^2 - (\sum_i o_i)^2} \sqrt{R \sum_i p_i^2 - (\sum_i p_i)^2}}$$

with $o_i, p_i \in \{0, 1\}$. The correlation coefficient for a whole dataset is then computed by averaging over the chains in that dataset. PCC values lie in the $[-1, 1]$ continuous interval, with 0 denoting complete uncorrelation, and ± 1 indicating direct and inverse perfect correlation, respectively. In our case, correlation values as closest to 1 as possible are desirable, meaning high similarity between observed and predicted data.

Relative Solvent Accessibility (RSA). Intuitively, the RSA value is an indicator of the percentage of the residue surface area that is exposed. Our tool computes RSA values of the residues in the hit sequences as follows. First, residue absolute Accessible Surface Area (ASA) values are retrieved form DSSP, then these values are normalized using Chothia method [29], i.e., ASA of a given residue X is divided by the maximum exposure area. We recall that the latter quantity is given by the ASA value of the same residue type in a Gly-X-Gly oligopeptide, with the side chain in a fully extended and standard conformation.

Exposure threshold. Our tool classifies residues into buried or exposed by means of the exposure threshold on RSA values associated to query residues. The threshold value is an input parameter.

In this work we set the default value of the exposure threshold to 20%. This default value has been used in [32] for the first time as the value that allows an even distribution of residues, with respect to solvent accessibility value, of the sequences in the considered dataset. This threshold value has often been considered as a reference in later works [4,5,7,8,12,13,14,16,18,19,20,22,23].

Similarity Depth (SD). In the first phase, our tool searches for sequences that are homologous to the query sequence using BLAST, and keeps the first hits to be processed later. We call *Similarity Depth* (SD) the number of hits selected by the tool at this stage, which is an input parameter of the tool.

The influence of SD on the overall performance has been studied empirically (with the exposure threshold at the default value). The observed results are shown in Figure 2.

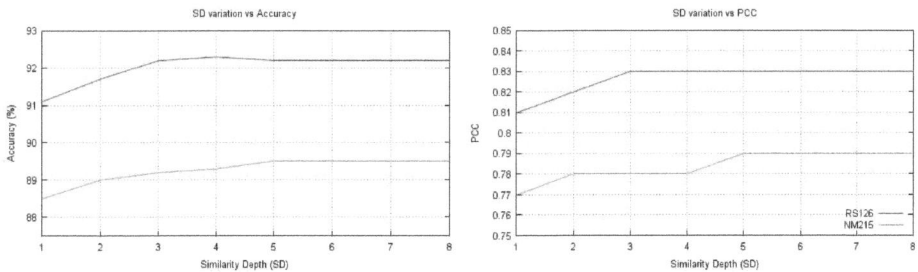

Fig. 2. Accuracy and PCC VS Similarity Depth value for the two datasets. The tool was run in default configuration (20% exposure threshold).

Observe that even with just one hit sequence (the most similar one), remarkable values of accuracy and correlation are obtained: 91.1% (resp. 88.5%) for accuracy and 0.81 (0.77) for correlation for the RS126 (resp. NM215) dataset. Increasing SD leads to a gradual refinement of prediction accuracy, apparently tending to a limit value for both accuracy and correlation when SD is greater than or equal to 5.

The above behavior could be explained by the following observation. Initially, the use of more hit sequences indeed helps to fill up the "higher-level" tables (2P2N being the highest), so that there is a clear improvement in the subsequent look-up phase. As the number of sequences considered further increases, either the pairwise-sequence identity keeps high, and thus the average values stored in the higher tables do not vary much, or otherwise the new RSA values go into lower-level tables, whose entries are likely not to be looked-up because of more probable higher-level hits.

These considerations led us to set SD to a default value of 5.

3.2   Results and Discussion

We compared our tool with some of the most representative and best performing RSA prediction tools available in literature. Competitors use very different approaches to predict the exposure state of residues. For each such approach we selected the best performing tool: [6] for the Information Theory (IT) approach, [8] for Probability Profiles (PP), SARpred [14] for Neural Networks (NN), RSA-PRP [20] for Support Vector Regression (SVR), [23] for a combination of Linear Regression and Support Vector Regression (LR+SVR), and SABLE [13] for a combination of Neural Networks and Linear Regression (NN+LR). We did not compare our tool against those that used a Real Values approach [33,21,15] (including the look-up table approach by Carugo *et al.* [5]), as these are not binary classifiers, which makes output comparison not straightforward. One might post-process Carugo *et al.* tool output using the same threshold used by our tool to produce a binary result, but the comparison might not result fair, as the tools were designed to address different problems.

Results are shown in Table 1: when a tool could not be downloaded or run properly, the reported results are taken from published papers. Missing entries are due to missing results in the original papers. Remember that the majority of tools do not take the exposure threshold as an input parameters, hence for some tests results are not available and we could not test all tools using all exposure threshold values. On the contrary, we ran our tool with any threshold value that has been used by other tool, always providing direct comparison.

The results show that our tool performs very well (in the considered datasets) both in terms of Correlation Coefficient and Accuracy, always outperforming the other tools where comparisons were possible. The obtained results are likely very close to the theoretical limit to solvent accessibility prediction, due to the intrinsic nature of variability for residues of proteins in their native state. In fact, RSA can reach about 10% of variability overall in protein chains with 100% of sequence identity [24].

Our tool is particularly reliable when the query sequence shows high similarity with known sequences, and less reliable otherwise. Nevertheless, our tool is positively

Table 1. Accuracy (and PCC) comparison with other methods and different threshold values on the NM215 and RS126 datasets. ^{c}Accuracy (and PCC) obtained by our tool on RS126 (resp. NM215) with exposure thresholds not used by other tools predicting RS126 (resp. NM215). Our tool is set to default configuration with $SD = 5$. For blank entries see the discussion in the text. ^{a} Only results for 4% and 9% threshold values are available, respectively. ^{b} Mattews' Correlation Coefficient (MCC) used, instead of PCC.

Tool/Approach (YEAR)	Exposure threshold			
NM215 dataset	**5%**	**10%**	**20%**	**25%**
IT (2001) [6]	75.1% (0.49)^a	75.9% (0.51)^a	-	74.4% (0.47)
PP (2003) [8]	75.7% (0.34)	73.4% (0.40)	-	71.6% (0.43)
SABLE/NN+LR (2004) [13]	76.8% (—)	77.5% (—)	77.9% (—)	77.6% (—)
SARpred/NN (2005) [14]	74.9% (0.31)^b	77.2% (0.50)^b	77.7% (0.56)^b	-
SVR+LR (2008) [23]	81.1% (0.68)	79.7% (0.68)	78.8% (0.68)	-
RSA-PRP/SVR (2010) [20]	77.1% (—)	77.0% (—)	77.5% (—)	77.4% (—)
Our Tool	**91.7% (0.78)**	**90.7% (0.79)**	**89.5% (0.79)**	**89.1% (0.78)**
Our Tool on RS126^c	**94.4% (0.78)**	**93.7% (0.80)**	**92.2% (0.83)**	**91.9% (0.83)**

RS126 dataset		9%	16%	23%	
IT (2001) [6]		78.2% (–)	77.5% (–)	77.4% (–)	
PP (2003) [8]		72.8% (0.39)	71.5% (0.42)	71.4% (0.43)	
Our Tool		**93.4% (0.80)**	**92.3% (0.81)**	**91.7% (0.82)**	
Our Tool on NM215^c		**90.9% (0.79)**	**90% (0.79)**	**89% (0.78)**	

affected by the continuous update of the PDB Data Bank: when new solved structures are added to the data bank, low performing query sequences might get better predictions if similar enough to the newly added ones.

The following examples clearly show how powerful our tool might be: our prediction for the protein chain 119LA [35] in the NM215 dataset (with default parameters) reaches accuracy 93% and correlation 0.86, compared to accuracy and correlation results, respectively, of 77% and 0.58 for SABLE [13], 83% and 0.62 for RSA-PRP [20], 80% and 0.56 for SARPRED [14]. Even better, our prediction for the protein chain 1bmv 1 [36] in the RS126 dataset (with default parameters) reaches accuracy 97% and correlation 0.95, compared to accuracy and correlation results, respectively, of 74% and 0.52 for SABLE, 70% and 0.42 for RSA-PRP, 69% and 0.37 for SARPRED.

With the aim of making a finer investigation of the good results obtained at the dataset level, we analyzed results also at single sequence level. Figure 3 and Figure 4 show the distributions of correlation values, respectively accuracy values, on the sequences composing the dataset. It can be seen that both values are clustered around the average: accuracy 92.2% and correlation 0.83 for RS126, accuracy 89.5% and correlation 0.79 for NM215.

The worst performing sequences bring accuracy down to 59% for RS126 and 57% for NM215, and correlation down to 0.19 and to 0.13, respectively. We deeply investigated the prediction process for low performing predictions and we found out that this happens mainly for one (or both) of the following reasons: (1) in the set of hit sequences

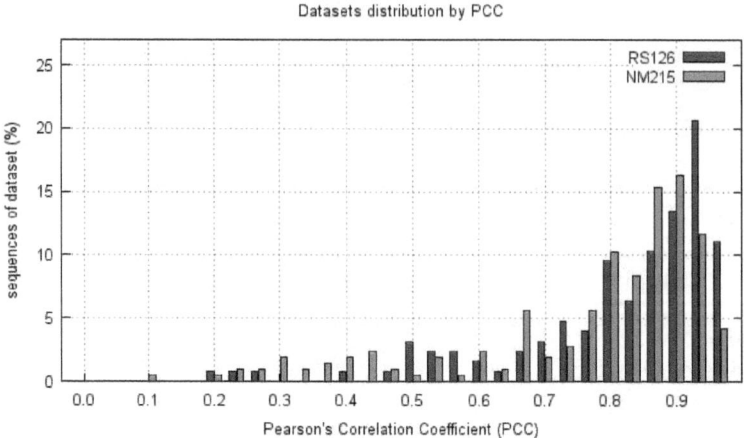

Fig. 3. PCC values reached in prediction, related with the percent number of sequences obtaining specific PCC value. Our tool was run in default configuration.

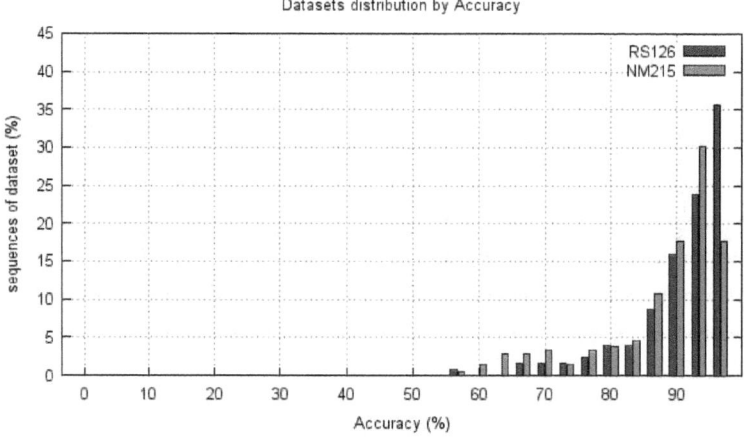

Fig. 4. Accuracy values reached in prediction, related with the percent number of sequences obtaining specific accuracy value. Our tool was run in default configuration.

there are sequences showing less than 30% of identity with the query and sharing local identity of at most three consecutive residues. This implies that the most reliable data for prediction, those in the 2P2N table, are completely absent, and hence that the prediction relies only on shorter contexts. (2) The set of hit sequences contains short sequences that do not cover the entire length of the query sequence; in this way, the prediction of uncovered portions of the query sequence is done according to data that refer to unaligned portions of the sequence.

The former problem is deeply connected with the approach adopted by our tool: if there is no solved structure similar enough to the sequence we wish to predict, then

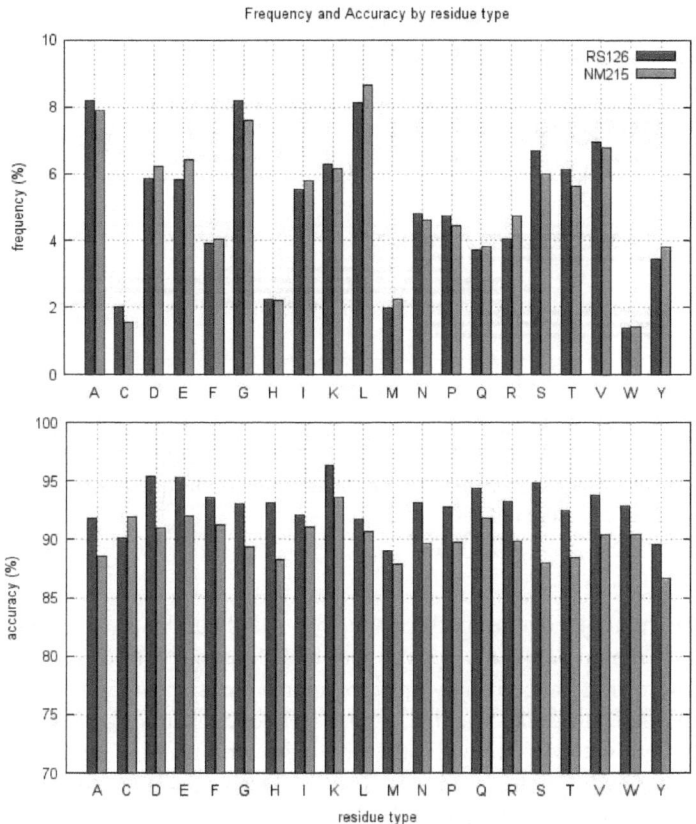

Fig. 5. Residue frequency (top) and accuracy (bottom). Frequency gives the number of times each type of residue appears in the dataset. Accuracy values refer to percent of correctly predicted exposure state for type of residue. Our tool was run in default configuration.

there is a small chance to return a reliable prediction. Nevertheless, the user might be advised of such a situation. On the other hand, the second problem can be somehow worked out (see Section 4 for some intuitions), but we leave this for the "work still to be done" agenda.

We also investigated the obtained results at residue level. Figure 5 shows frequency and prediction accuracy distributions among standard amino acids. Observe that frequency distribution of residues is quite conserved among the two datasets, allowing us to make comparisons between them.

The first and probably most important observation is that the range of accuracy prediction distribution is reasonably small, being about 7%. A finer look reveals that the worst and best predicted residues in both datasets are M (Met) and K (Lys), respectively. Note, however, that even M exhibits sufficiently high accuracies (namely 87% and 89% in NM215 and RS126, respectively), while K reaches such very good figures as 93.5% and 96.5% in NM215 and RS126, respectively.

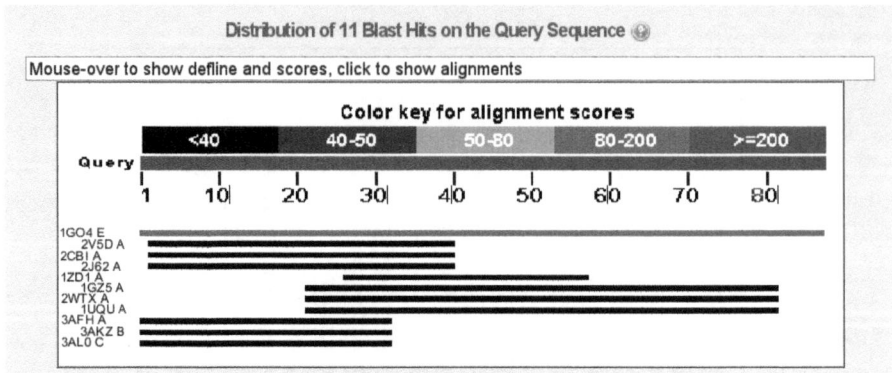

Fig. 6. BLAST output showing the coverage of the most similar sequences to the 1GO4 E query. Prediction using 2V5D A and 1GZ5 A, instead of 2V5D A and 2CBI A, leads to a significant improvement, as the former sequences span the whole length of the query sequence. Our tool was run with SD = 2, and exposure threshold of 20%.

To improve the tool's overall performance, we should address our attention to those low-performing amino acids that appear with high frequency. In this respect, M itself is not very interesting, since it only appears approximately twice every 100 residues. One such candidate is instead Alanine, whose frequency is among the highest (around 8%). The problem with A (like with other low-performing amino acids) is that it does not have a strong hydrophobic nor hydrophilic preference, and its exposure state floats between buried and exposed depending on the surrounding local environment. It is thus clear that for A's accuracy to improve more context information is desirable. Note that this behavior of A is in agreement with the already mentioned PCA study [31].

Other effects that can be noticed in some residue behavior are probably due to the mixing influence of the two problems (i.e., low local sequence-identity in the hit sequences in the neighborhoods of the considered residue, and low query sequence coverage) that we mentioned when discussing the results at the sequence level. In particular, we may notice that for some amino acid the ranked performance is completely reversed in the two datasets. This is the case of T (Thr) and S (Ser): these are among the best predicted for the RS126 dataset (92.5% and 95% accuracy, respectively), and among the worst ones for the NM215 dataset (around 87%). On the other hand, amino acid C (Cys) is one of the best predicted for the NM215 dataset (with an accuracy of 92%) and one of the worst for the RS126 dataset (accuracy 90%).

Our last investigation deals with look-up tables statistics. As it might be expected, 2P2N is generally a very sparse matrix (on the average, no more than 0.1% of the cells contain a non zero value, for both datasets), nevertheless hits do occur frequently during the prediction process: 84.3% of the times the tool finds a hit in the 2P2N table, for the NM215 dataset, and 93.9% of the times for RS126. Table 1P1N is clearly less sparse than 2P2N, with 3.9% (resp. 2,5%) of non zero entries for the NM215 (resp. RS126) dataset, but its contents are used only around twice to four times, depending on the dataset, for each 100 table look-ups. Finally, the most populated tables are 1P and 1N, which together have 43.3% and 32.8% of non zero entries in NM215 and RS126, respectively, and are accessed from 3 to 10 times every 100 look-ups, depending on the dataset. For the sake of completeness, we also mention that 1.4% of the times the exposure states ate recovered from the default PCA values.

This data clearly show how 2P2N function is truly relevant, since it stores information for the largest context (of the central residue), and the most similar replication of the environment surrounding each amino acid. Should a stretch of the query sequence match a 5-residues pattern in the tables, this would be a very close replica of the former one, hence representing a very similar peptide environment and providing a reliable prediction. Indeed, we observed that predictions done with very few hits in 2P2N are not very reliable predictions, and vice-versa. This also suggests that the main avenue for a further improvement is not an increase of the context size, but rather an increase in the number of hits in 2P2N.

4   Conclusions and Future Work

In this paper we described a tool that is able to produce very reliable predictions on the exposed/buried state of protein amino acids. The tool bases its predictions mainly on sequence homology, by using information of already solved protein structures that show some degree of similarity with the sequence under prediction. Results obtained on consolidated benchmarks show that our tool clearly outperforms existing tools adopting alternative strategies.

Although the results obtained up to now are extremely encouraging, there still is enough room for further analysis and possible improvements. First of all, the tools high performances should be confirmed (or re-assessed) on larger datasets containing chains that have been solved more recently. Secondly, additional work must be done to address some of the problems discussed in the previous section and get possibly even better results.

As for the latter point, we plan to address at least the following two issues.

(1) As we pointed out in section 3.2, our predictions are less reliable when the set of hit sequences does not cover the entire query sequence; *i.e.*, there are large enough portions of the query sequence that are not aligned with any portions of the hit sequences. A major optimization of the tool would be to select, in the output returned by BLAST, a set of sequences that covers the entire length of the query sequence, while maintaining a high similarity level.

Figure 6 shows an example where the sequence with PDB identifier 1GO4 E [37] is only partially covered by the best scoring sequences found by BLAST. If we run the current version of the tool, considering only the the two best scoring similar sequences obtained by a BLAST search (namely 2V5D A and 2CBI A), we achieve 62% accuracy and 0.18 correlation in prediction, while a better result of 70% accuracy and 0.40 correlation is achieved if the prediction is done by using the two most similar sequences that span the whole length of the query sequence (namely 2V5D A and 1GZ5 A) [38,39,40]. Although the use of 1GZ5 A in place of 2CBI A gives only a moderate improvement in accuracy, this example does suggest that it is possible to achieve better predictions by improving the query coverage (yet simultaneously storing statistically relevant values in the 2P2N table).

Conversely, neglecting "outlying" stretches of hit sequences much longer than the query, which only share a high local similarity with it, might also facilitate data reliability and noise reduction in prediction, because RSA values relative to the residues in those portions of the sequences will not affect look-up table entries.

(2) In its current version, the tool makes predictions by accessing look-up tables by means of the exact context surrounding the residue under consideration. It might be interesting to investigate the possibility of accessing tables using similar, but possibly not equal, contexts. Here "similar" means that we allow the substitution of some (one or two) context residues with others that do not significantly alter its neighbor exposure state. Clearly, the choice of appropriate substitution matrices is crucial here, but the payoff could be an increase in the number of hits in the highest 2P2N look-up table, with the already pointed out benefits on the performance.

Acknowledgments. Authors wish to thank Prof. M.C. Menziani for the useful discussions during the research process that led to the present work.

References

1. Jones, S., Thornton, J.M.: Analysis of Protein-Protein Interaction Sites Using Surface Patches. J. Mol. Biol. 272, 132–143 (1997) 138
2. Wako, H., Blundell, T.L.: Use of Amino Acid Environment-Dependent Substitution Tables and Conformational Propensities in Structure Prediction from Aligned Sequences of Homologous Proteins. I. Solvent accessibility classes. J. Mol. Biol. 238, 682–692 (1994) 138
3. Chakrabarti, P., Janin, J.: Dissecting Protein-Protein Recognition Sites. Proteins 47, 334–343 (2002) 138
4. Rost, B., Sander, C.: Conservation and Prediction of Solvent Accessibility in Protein Families. Proteins 20, 216–226 (1994) 139, 143
5. Carugo, O.: Predicting Residue Solvent Accessibility From Protein Sequence by Considering the Sequence Environment. Protein Eng. 13, 607–609 (2000) 139, 141, 143, 144
6. Naderi-Manesh, H., Sadeghi, M., Arab, S., Moosavi Movahedi, A.A.: Prediction of Protein Surface Accessibility with Information Theory. Proteins 42, 452–459 (2001) 139, 142, 144, 145
7. Thompson, M.J., Goldstein, R.A.: Predicting Solvent Accessibility: Higher Accuracy Using Bayesian Statistics and Optimized Residue Substitution Classes. Proteins 25, 38–47 (1996) 139, 143
8. Gianese, G., Bossa, F., Pascarella, S.: Improvement in Prediction of Solvent Accessibility by Probability Profiles. Protein Eng. 16, 987–992 (2003) 139, 143, 144, 145
9. Holbrook, S.R., Muskal, S.M., Kim, S.H.: Predicting Surface Exposure of Amino Acids from Protein Sequences. Protein Eng. 3, 659–665 (1990) 139
10. Rost, B., Sander, C.: Combining Evolutionary Information and Neural Networks to Predict Protein Secondary Structure. Proteins 19, 55–72 (1994) 142
11. Ahmad, S., Gromiha, M.M.: NETASA: Neural Network Based Prediction of Solvent Accessibility. Bioinformatics 18, 819–824 (2002) 139
12. Pollastri, G., Baldi, P., Fariselli, P., Casadio, R.: Prediction of Coordination Number and Relative Solvent Accessibility in Proteins. Proteins 47, 142–153 (2002) 139, 143

13. Adamczak, R., Porollo, A., Meller, J.: Accurate Prediction of Solvent Accessibility Using Neural Networks Based Regression. Proteins 56, 753–767 (2004) 139, 143, 144, 145
14. Garg, A., Kaur, H., Raghava, G.P.S.: Real Value Prediction of Solvent Accessibility in Proteins Using Multiple Sequence Alignment and Secondary Structure. Proteins 61, 318–324 (2005) 139, 143, 144, 145
15. Dor, O., Zhou, Y.: Real-SPINE: An Integrated System of Neural Networks for Real-value Prediction of Protein Structural Properties. Proteins 68, 76–81 (2007) 139, 144
16. Li, X., Pan, X.M.: New Method for Accurate Prediction of Aolvent Accessibility from Protein Sequence. Proteins 42, 1–5 (2001) 139, 143
17. Wang, J., Lee, H., Ahmad, S.: Prediction and Evolutionary Information Analysis of Protein Solvent Accessibility Using Multiple Linear Regression. Proteins 61, 481–491 (2005) 139
18. Yuan, Z., Burrage, K., Mattick, J.S.: Prediction of Protein Solvent Accessibility Using Support Vector Machines. Proteins 48, 566–570 (2002) 139, 143
19. Nguyen, M., Rajapakse, J.: Prediction of Protein Relative Solvent Accessibility with a two-stage SVM Approach. Proteins 59, 30–37 (2005) 139, 143
20. Meshkin, A., Ghafuri, H.: Prediction of Relative Solvent Accesibility by Support Vector Regression and Best-First Method. EXCLI Journal 9, 29–38 (2010) 139, 143, 144, 145
21. Wang, J.-Y., Ahmad, S., Gromiha, M.M., Sarai, A.: Look-up Tables for Protein Solvent Accessibility Prediction and Nearest Neighbor Effect Analysis. Biopolymers 75, 209–216 (2004) 139, 141, 144
22. Chen, H., Zhou, H.X.: Prediction of Solvent Accessibility and Sites of Deleterious Mutations from Protein Sequence. Nucleic Acids Res. 33, 3193–3199 (2005) 139, 143
23. Chen, K., Kurgan, M., Kurgan, L.: Sequence Based Prediction of Relative Solvent Accessibility Using two-stage Support Vector Regression with Confidence Values. J. Biomed. Sci. Eng. 1, 1–9 (2008) 139, 143, 144, 145
24. Flores, T.P., Orengo, C.A., Moss, D.S., Thornton, J.M.: Comparison of Conformational Characteristics in Structurally Similar Protein Pairs. Protein Sci. 2, 1811–1826 (1993) 139, 144
25. Cuff, J.A., Barton, G.J.: Application of Multiple Sequence Alignments Profiles to Improve Protein Secondary Structure Prediction. Proteins 40, 502–511 (2000) 139
26. Berman, H.M., Westbrook, J., Feng, Z., Gilliland, G., Bhat, T.N., Weissig, H., Shindyalov, I.N., Bourne, P.E.: The Protein Data Bank. Nucleic Acids Res. 28, 235–242 (2000) 139
27. Altschul, S.F., Gish, W., Miller, W., Myers, E.W., Lipman, D.J.: Basic Local Alignment Search Tool. J. Mol. Biol. 215, 403–410 (1990) 139
28. Kabsch, W., Sander, C.: Dictionary of Protein Secondary Structure: Pattern Recognition of Hydrogen-Bonded and Geometrical Features. Biopolymers 22, 2577–2637 (1983) 139
29. Chothia, C.: The Nature of the Accessible and Buried Surfaces in Proteins. J. Mol. Biol. 105, 1–12 (1976) 143
30. Carugo, O.: Prediction of Polypeptide Fragments Exposed to the Solvent. Silico Biology 3, 35 (2003) 141
31. Palmieri, L., Federico, M., Leoncini, M., Montangero, M.: Sequence-Based Prediction of Solvent Accessibility in Proteins. University of Modena and Reggio Emilia, M2CSC doctoral research school, internal report (2009) 142, 148
32. Rose, G.D., Geselowitz, A.R., Lesser, G.J., Lee, R.H., Zehfus, M.H.: Hydrophobicity of Amino Acid Residues in Globular Proteins. Science 229, 834–838 (1985) 143
33. Ahmad, S., Gromiha, M.M., Sarai, A.: Real Value Prediction of Solvent Accessibility from Amino Acid Sequence. Proteins 50, 629–635 (2003) 144
34. Brenner, S.E., Chothia, C., Hubbard, T.J.P.: PNAS 95, 6073–6078 (1998)
35. Blaber, M., Lindstrom, J.D., Gassner, N., Xu, J., Heinz, D.W., Matthews, B.W.: Energetic Cost and Structural Consequences of Burying a Hydroxyl Group within the Core of a Protein Determined from Ala–>Ser and Val–>Thr Substitutions in T4 lysozyme. Biochemistry 32, 11363–11373 (1993) 145

36. Chen, Z.G., Stauffacher, C., Li, Y., Schmidt, T., Bomu, W., Kamer, G., Shanks, M., Lomonos-soff, G., Johnson, J.E.: Protein-RNA Interactions in an Icosahedral Virus at 3.0 A Resolution. Science 245, 154–159 (1998) 145
37. Sironi, L., Mapelli, M., Knapp, S., Antoni, A., Jeang, K.T., Musacchio, A.: Crystal Structure of the Tetrameric Mad1-Mad2 Core Complex: Implications of a 'Safety Belt' Binding Mechanism for the Spindle Checkpoint. Embo. J. 21, 2496 (2002) 149
38. Ficko-Blean, E., Gregg, K.J., Adams, J.J., Hehemann, J.H., Smith, S.J., Czjzek, M., Boras-ton, A.B.: Portrait of an Enzyme, a Complete Structural Analysis of a Multimodular beta-N-acetylglucosaminidase from Clostridium Perfringens. J. Biol. Chem. 284, 9876–9884 (2009) 149
39. Rao, F.V., Dorfmueller, H.C., Villa, F., Allwood, M., Eggleston, I.M., Van Aalten, D.M.F.: Structural Insights into the Mechanism and Inhibition of Eukaryotic O-GlcNAc Hydrolysis. Embo. J. 25, 1569 (2006) 149
40. Gibson, R.P., Turkenburg, J.P., Charnock, S.J., Lloyd, R., Davies, G.J.: Insights into Tre-halose Synthesis Provided by the Structure of the Retaining Glucosyltransferase OtsA. Chem. Biol. 9, 1337 (2002) 149

Removing Artifacts of Approximated Motifs[*]

Maria Federico[1,2] and Nadia Pisanti[2]

[1] Dipartimento di Ingegneria dell'Informazione, Università di Modena e Reggio
Emilia, Italy
[2] Dipartimento di Informatica, Università di Pisa, Italy

Abstract. Frequent patterns (motifs) in biological sequences are good
candidates to correspond to structural or functional important elements.
The typical output of existing tools for the exhaustive detection of approximated motifs is a long list of motifs containing some real motifs (i.e.,
patterns representing functional elements) along with a large number of
random variations of them, called *artifacts*. Artifacts increase the output
size, often leading to redundant and poorly usable results for biologists.
In this paper, we provide a new solution to the problem of separating
real motifs from artifacts. We define a notion of motif maximality, called
maximality in conservation, which, if applied to the output of existing
motif finding tools, allows us to identify and remove artifacts. Their detection is based on the fact that variations of a motif share a large subset
of occurrences of the real motif, but the latter is more conserved than
any of its artifacts. Experiments show that the tool we implemented according to such definition allows a sensible reduction of the output size
removing artifacts with a negligible time cost.

Keywords: Maximal Motifs, Motif conservation, Biological Sequences.

1 Introduction

Frequent patterns (motifs) in biological sequences usually correspond to functional or structural important elements. Motifs are patterns that occur more
often than expected in a biological sequence, or which are surprisingly shared
by several distinct sequences, that is motifs are patterns which are statistically
over-represented [1]. Motif extraction tools aim at inferring these patterns. In
applications involving biological sequences, given the frequency of sequencing
errors as well as point mutations, the inferred motifs are *approximated*, that is,
distinct occurrences of the same motif are not necessarily identical, but just *similar*, according to a given similarity notion. From the computational complexity
point of view, this makes the task of finding over-represented patterns harder,
whatever is the type of approximation one uses. The *Hamming distance* is defined between patterns of the same length and it simply consists of the number
of differences that occur between them. One usually sets a maximum allowed

---

[*] This work was supported in part by MIUR of Italy under project AlgoDEEP prot.
2008TFBWL4.

C. Böhm et al. (Eds.): ITBAM 2011, LNCS 6865, pp. 153–167, 2011.

distance and then requires that the motifs differ by at most that number of letters substitutions.

The typical output of existing tools for the exhaustive detection of motifs is a long list of motifs containing some real motifs (i.e., patterns representing functional elements) along with a large number of random variations of them (called *artifacts* in [1]). Artifacts are detected as they satisfy frequency and conservation requirements of the motifs, but they actually do so because the number of their occurrences is "artificially" incremented by the presence of a real motif; the reason is that if a real motif is over-represented, its artifacts are over-represented as well. Artifacts enlarge the output size, and for this reason they often lead to a poor usability of the results that are too large and noisy to be investigated by biologists. Motif finding algorithms should reach a trade-off between good soundness (*i.e.*, they output very few motifs that are very likely to represent functional elements) and completeness (*i.e.*, they loose few or no motifs representing real elements). However, many existing methods are designed to give a complete list of statistically over-represented patterns in order to maximize their completeness. The downside of this choice is that the list will also contain typically hundreds of artifacts, as a consequence they have a very low soundness. Some motif finding algorithms solve the problem of separating artifacts from real motifs by sorting motifs in the output based on their statistical significance, such that real motifs should be at highest positions of the list. Typical measures of statistical significance of motifs are Z-*score*, p-*value* and X^2-*value*. These measures provide, for each pattern, a score based on the ratio between the *observed* number of motif occurrences in input sequences and their *expected* number in random sequences or in reference sequences (for example, *background* sequences which do not contain any functional element, and *shuffled* sequences generated by shuffling input sequences but preserving the exact same frequency of bases or amino acids). A survey of these methods can be found in [5]. The statistical significance measures sometimes lead the motif extraction toward more interesting solutions, but they are often computed during a post-processing step, after motifs are found, in order to sort them. Some methods group high similar motifs to improve the readability of output, but similarity alone may not be a reliable criteria. Another approach is to iteratively search for the most significant motif and to mask its occurrences in input sequences, such that none of its random variations is found in successive iterations. Of course, applying this technique all overlapped motifs are lost.

In this paper, we provide an alternative solution to the problem of separating real motifs from artifacts. We define a notion of motif maximality, called *maximality in conservation*, which, if applied to the output of existing motif finding tools, allows us to identify and remove artifacts from the output based on the fact that variations of a motif share the same or a subset of occurrences of the real motif but the latter is more conserved, that is it has less differences with its occurrences, than the artifacts.

The notion of maximality in conservation, applying to motifs of the same length, is something different from other existing notions of maximality in specificity and extension (which apply to motifs of different length) defined for motifs approximated with Hamming distance [2,3,6], edit distance [7], don't care [10,11,4] and using a degenerate alphabet [12]. On the other hand, notions of maximality in extension and our notion of maximality in conservation can coexist in order to reduce the redundancy of output provided by motif finding tools given a minimum motif length.

Experiments (see Section 5) show that the tool we implemented according to our definition of maximality in conservation allows a sensible reduction of the output size removing artifacts with a negligible time cost.

2   Preliminary Definitions

We consider strings that are finite sequences of characters drawn from an alphabet Σ. In particular we will focus out attention on the DNA alphabet $\Sigma = \{A, C, G, T\}$, and hence when a string represents a DNA sequence. We denote by $s[i]$ the character at position i in a string s and by $|s|$ the length of s. Consecutive characters of s form a *substring* of s. The substring of s that starts from position i and ends at position j is denoted by $s[i..j]$, where $1 \leq i \leq j \leq |s|$. Given a string x drawn from the same alphabet as s (or from one of its subsets), we say that $s[i..j]$ *exactly occurs* at position i in s if and only if $x = s[i, i+|x|-1]$. In this case, we also say that $s[i, i + |x| - 1]$ is an *occurrence* of x in s. The Hamming distance between two strings of the same length x and y, denoted as $d_H(x, y)$, is the smallest number of letter substitutions that transform x into y (or vice versa as the distance is symmetric). Given an integer $e \geq 0$, we say that a substring y of a string s is an *e-occurrence* of a string x, if and only if $d_H(x, y) \leq e$. In this case we will also talk about an *approximate occurrence*, or simply an *occurrence*, of x in s. The list of all occurences of a pattern x in s is denoted by $L_{(e,x)}$ and is called *position set*. We will also make use of the *occurrence set*, denoted by $O_{(e,x)}$, that contains couples of integers (p_i, d_i), where $p_i \in L_{(e,x)}$ and $d_i = d_H(x, s[p_i..p_i + |x| - 1])$ for each $1 \leq i \leq |O_{(e,m)}|$. Clearly, $|L_{(e,m)}| = |O_{(e,m)}|$ and $d_i \leq e$ for each $1 \leq i \leq |O_{(e,m)}|$.

Definition 1. *Given a sequence s, a quorum $q \geq 2$, and $e \geq 0$, a pattern m is a motif iff $|L_{(e,m)}| \geq q$.*

If $e = 0$ we speak about *exact motifs*, because no differences between motifs and their occurrences are allowed; otherwise, when $e > 0$, we call them *approximate motifs*. The traditional motif extraction problem gives as input: (i) the string in which one wants to find the repeated motifs (or the set of strings in which one wants to find the common motifs); (ii) the quorum; (iii) the (minimal) length ℓ required for the motifs; (iv) optionally, an approximation measure (e.g. the Hamming distance), and the value of e for the approximation measure. The requested output is typically the set of all patterns of length (at least) ℓ that have at least q (possibly approximated) occurrences in s, that is, the complete set of motifs. Within this traditional framework, the complete output can be very noisy as it contains redundant data. Indeed, once that a length is fixed, several motifs may share the same occurrence set or, more frequently, a (possibly very large) subset of it. Among all motifs having "almost" the same set of occurrences, it is likely that just one of them represents a real functional element, while the others are the artifacts. In a sequence-based selection framework, the best candidate to be the real functional element is the most conserved motif, while the artifacts are very similar to it and mostly less conserved. In other words, the Hamming distance between the real motif and its real occurrences is on average lower than that of an artifact. We call such motifs the *maximally conserved motifs*. In this paper, we suggest a way to output only maximally conserved motifs according to a notion of maximality in conservation that we introduce here; we will apply it to the output of motif extraction tools in a post-processing step. To this purpose, we first introduce the notions of sum and mean of Hamming distances between a motif and its occurrences.

Definition 2. *Given a sequence s or a set S of sequences and an integer $e \geq 0$. Let $(p_1, d_1), \ldots, (p_t, d_t)$ be elements of the occurrence set $O_{(e,x)}$ of pattern x. We define:*

$$S_x = \sum_{i=1}^{|O_{(e,x)}|} d_i$$

and

$$\mu_x = \frac{S_x}{|O_{(e,x)}|}$$

the sum *and the* mean *of the Hamming distances between the pattern x and its occurrences in s or S, respectively.*

Before defining the notion of maximality in conservation for approximate motifs, we introduce the notion of (ζ, δ)-conservation of a motif with respect to another one. Intuitively, if a motif is (ζ, δ)-conserved with respect to another one, then the first is somehow more significant in terms of a suitable combination of frequency and conservation. More formally:

Definition 3. *Given a sequence s or a set S of sequences and integers $e \geq 0$, $\zeta > 0$ and $\delta > 0$. Let m and m' be two approximate motifs of length l, and let $L_{(e,m)}$ and $L_{(e,m')}$ be their position sets, respectively, such that $|L_{(e,m)} \cap L_{(e,m')}| \geq \zeta$.*

The motif m is (ζ, δ)-conserved with respect to m' iff one of the three following cases holds:

1. *$|L_{(e,m)}| = |L_{(e,m')}|$ and $S_m < S_{m'}$,*
2. *$|L_{(e,m)}| > |L_{(e,m')}|$ and $\mu_m \leq \mu_{m'}$,*
3. *$|L_{(e,m)}| < |L_{(e,m')}|$ and $\mu_m \leq \mu_{m'} - \delta$.*

The suitable values for thresholds ζ and δ are dependent on the particular biological application, and in particular, on the input sequences and on the kind of functional elements to be sought in them. What we will say in this paper about these parameters has general purposes and it is not intended to any specific application.

The threshold ζ provides a lower bound to the number of occurrences that two motifs should share so that the (ζ, δ)-conservation of a motif with respect to another one has sense, based on the Definition 3. Indeed, it is obvious that the lower is the threshold ζ, the higher will be the probability to find two motifs satisfying one of the three conditions in Definition 3. On the other hand, the (ζ, δ)-conservation of a motif with respect to another one implies that the two motifs represent, with a different degree of accuracy (higher for the (ζ, δ)-conserved motif, lower for the other), the same functional element, and therefore it is useless to keep both because the (ζ, δ)-conserved motif contain all the information needed to represent the other motif as well. If the two motifs share very few occurrences, this theory does not make sense because we compare two motifs which are very different and therefore they are unlikely to represent the same functional element.

Let us see a simple synthetic example of application of the notion of (ζ, δ)-conservation that allows to distinguish real motifs from artifacts.

Example 1. Assume to look for motifs of length 3 with $q = 2$ and $e = 1$ in the input sequence $s = ACCTACGACC$. The pattern $m_1 = ACC$ with occurrence set $O_{(1,m_1)} = \{(1,0), (5,1), (8,0)\}$ is a motif, but we find also motifs $m_2 = ACA$ with occurrence set $O_{(1,m_2)} = \{(1,1), (5,1), (8,1)\}$, $m_3 = ACG$ with $O_{(1,m_3)} = \{(1,1), (5,0), (8,1)\}$, $m_4 = ACT$ with $O_{(1,m_4)} = \{(1,1), (2,1), (5,1), (8,0)\}$, $m_5 = AGC$ with $O_{(1,m_5)} = \{(1,1), (8,1)\}$, $m_6 = ATC$ with $O_{(1,m_6)} = \{(1,1), (8,1)\}$, $m_7 = CCC$ with $O_{(1,m_7)} = \{(1,1), (2,1), (8,1)\}$, $m_8 = GCC$ with $O_{(1,m_8)} = \{(1,1), (7,1), (8,1)\}$, $m_9 = TCC$ with $O_{(1,m_9)} = \{(1,1), (4,1), (8,1)\}$.

The output consisting of all the motifs $m_1, m_2, m_3, m_4, m_5, m_6, m_7, m_8, m_9$ is quite noisy in that it contains nine motifs whose occurrences lists heavily overlap, and therefore there is a clear redundancy that should be removed. Setting $\zeta = 2$ and $\delta = 1$ we have that:

- The motif m_1 is $(2,1)$-conserved with respect to m_2, m_3, m_7, m_8, m_9 because:
  - $|L_{(1,m_1)} \cap L_{(1,m_j)}| = 3$ for $j = 2, 3$ and $|L_{(1,m_1)} \cap L_{(1,m_k)}| = 2$ for $k = 7, 8, 9$;
  - $|L_{(1,m_1)}| = |L_{(1,m_i)}| = 3$ for $i = 2, 3, 7, 8, 9$;
  - $S_{m_1} < S_{m_i}$ per $i = 2, 3, 7, 8, 9$, indeed $S_{m_1} = 1$, $S_{m_j} = 3$ for $j = 2, 7, 8, 9$ and $S_{m_3} = 2$.

  Hence, case 1 of Definition 3 holds.
- The motif m_1 is $(2,1)$-conserved with respect to m_5, m_6 because:
  - $|L_{(1,m_1)} \cap L_{(1,m_i)}| = 2$ for $i = 5, 6$;
  - $|L_{(1,m_1)}| = 3 > |L_{(1,m_i)}| = 2$ for $i = 5, 6$;
  - $\mu_{m_1} = 1/3 < \mu_{m_i} = 1$ for $i = 5, 6$.

  Hence, case 2 of Definition 3 holds.
- The motif m_1 is $(2,1)$-conserved with respect to m_4 because:
  - $|L_{(1,m_1)} \cap L_{(1,m_4)}| = 3$;
  - $|L_{(1,m_1)}| = 3 < |L_{(1,m_4)}| = 4$;

- $\mu_{m_1} < \mu_{m_4} - \delta$, indeed $\mu_{m_1} = 1/3$ and $\mu_{m_4} = 1$.

Hence, case 3 of Definition 3 holds.

Notice that no other motif is $(2,1)$-conserved. The notion of maximality we would like to define should result in acknowledging that $m_1 = ACC$ is the real motif, whereas the others are its artifacts. Observe that a selection purely and roughly based on frequency would have instead focused on m_4.

Clearly, the notion of (ζ, δ)-conservation provides just an heuristics which in some cases does not allow us to discern between two motifs, because it may be that none of them is (ζ, δ)-conserved based on the Definition 3. Indeed, there exist cases in which, even though two motifs share many occurrences, the mean of Hamming distances between the two motifs and their occurrences is the same, as shown in Example 2. In the same way, it may happen that the mean of Hamming distances between the less frequent motif and its occurrences is smaller than the ones of the most frequent motif but the difference is not large enough to determine the (ζ, δ)-conservation of the former (this is controlled by the threshold δ), as Example 3 shows.

Example 2. Consider the input sequence $s = AACAAGAACAAGA$ and assume $e = 1$ and $q \geq 2$. The patterns $m = AAC$ and $m' = AAG$ are both motifs with occurrence sets $O_{(1,m)} = \{(1,0),(4,1),(7,0),(10,1)\}$ and $O_{(1,m')} = \{(1,1),(4,0),(7,1),(10,0)\}$, respectively. The two motifs m and m' occur at the same positions and the sum of Hamming distances between each motif and its occurrences is 2 for both motifs, therefore it does not make sense to speak about (ζ, δ)-conservation of a motif with respect to the other.

Example 3. Consider the input sequence $s = AAGAACAACAAGAG$ and assume $e = 1$ and $q \geq 2$. The pattern $m = AAC$ having occurrence set $O_{(1,m)} = \{(1,1),(4,0),(7,0),(10,1)\}$ is a motif for which $\mu_m = 1/2$. The pattern $m' = AAG$ with occurrence set $O_{(1,m')} = \{(1,0),(4,1),(7,0),(10,1),(12,1)\}$ is a motif as well and $\mu_{m'} = 3/5$ ($|O_{(1,m)}| < |O_{(1,m')}|$). The two motifs m and m' share 4 occurrences and $\mu_{m'} = 3/5 > \mu_m = 1/2$, therefore $\mu_{m'} - \mu_m = 1/10$. If we fix $\delta \geq 0.30$, based on case 3 of Definition 3 none of m and m' is (ζ, δ)-conserved with respect to the other.

In Section 5 we will see how to fix parameters ζ and δ such that motifs representing real functional elements are (ζ, δ)-conserved with respect to their artifacts.

We define the maximality in conservation of a motif as follows:

Definition 4. *Given two integers $\zeta > 0$ and $\delta > 0$. An approximate motif m of length l is* maximally conserved *iff for each approximate motif m_i of length l such that $|L_{(e,m)} \cap L_{(e,m_i)}| \geq \zeta$, m is (ζ, δ)-conserved with respect to m_i.*

Based on this definition, the motif $m_1 = ACC$ in Example 1 would result to be the only maximally conserved one.

3   Maximality in Conservation versus Measures of Statistical Significance

The notion of maximality in conservation for approximate motifs proposed in Section 2 is completely different from the measures of statistical significance proposed in literature to separate real motifs from artifacts, such as *Z-score*, X^2-*value* and *p-value*. These measures, as anticipated in Section 1, provide for each motif a score based on the *observed* amount of motif occurrences in input sequences with respect to their *expected* number in random sequences. A problem which may raise using these measures is that, sorting the list of motifs provided by the extraction tools based on their statistical significance, in some cases artifacts are at higher positions in the list with respect to the real motif. In [1], authors say that in a sequence s a motif m "*explains*" a motif m' if the number of occurrences of m' is not significantly larger than that expected when we factor in our knowledge of where m occurs in s. Thus, if m explains m', the occurrences of m' can be interpreted as mere chance occurrences. If m represents a functional element, reporting m' would be unnecessary, and possibly misleading. Consider, for example, a sequence where the motif $ACGCCT$ occurs 80 times, though it was expected to occur only 20 times. Consider also $ACGCCA$, expected to occur about 10 times, and occurring 40 times. Treated seperately, both of these might be considered to be over-represented. However, if we take into consideration the fact that $ACGCCT$ occurs 80 times, we shall expect $ACGCCA$ to occur about 40 times, and so the former explains the latter. Based on this observation, in [1] authors address the problem of typical statistical significance metrics described above by introducing the *conditional Z-score* of m, given $e_1, \ldots, e_k$: $Z(m|e_1, \ldots, e_k) = (N_m - \mu_m)/\sigma_m$, where N_m is the number of occurrences of a motif m given the occurrences of motifs $e_1, \ldots, e_k$, μ_m is the number of expected occurrences of m and σ_m is the variance of N_m, which provides a measure of the *explanation* of a motif. A high value of this statistical measure implies that the motif m occurs more often than expected by chance, also taking into account the occurrences of motifs $e_1, \ldots, e_k$.

The notion of maximality in conservation proposed in this paper for approximate motifs provides an alternative method to separate real motifs from their artifacts. This notion is based on the fact that an artifact shares a large subset of occurrences with the real motif, but the real motif is more conserved, that is, it underwent a smaller number of mutations in the evolution (this follows from the fact that the mean of the Hamming distances between the motif and its occurrences is less than the one of the artifacts with respect to their occurrences), which is a possible indication of its functional importance. This is the main motivation of our choice for the notion of maximal conservation for motifs. Therefore, the maximality in conservation of a motif, should not just be linked to its frequency in biological sequences with respect to the one in random sequences (which is measured by the metrics of the statistical significance), but it should rely on a more general view including also the comparison between motifs which are present in input sequences and have almost the same occurrences, in order to determine the most conserved motifs.

In the next section, we present a post-processing filter to remove artifacts from the output of motif extraction tools and therefore to keep maximally conserved motifs only.

4   Post-processing Filter to Check Maximality in Conservation of Motifs

In this section we will present a post-processing filter, based on the maximal conservation, that we developed to remove artifacts from the output of tools for the extraction of approximate motifs.

The naive solution to this problem consists in computing the intersection between the occurrence set of each motif with the occurrence set of all the other motifs, of the same length and lexicographically sorted, in the list of motifs returned by extraction tools. For all pairs of motifs such that the size of the intersection is greater than the parameter ζ (that may be two artifacts of the same motif, or a real motif and one of its artifacts), the algorithm checks the conditions of Definition 3 to determine which is the (ζ, δ)-conserved motif, and removes the other from the list of motifs.

Our filter provides an alternative solution to the one described above which, as we will see in Section 5, is very efficient in practice, and that works as follows.

The inefficiency of the naive solution principally depends on the fact that the intersection between occurrence sets of two motifs is computed also for several motifs which do not share any occurrence. To address this problem, our filter uses an array A idexed by the positions of the input sequences, in addition to the list L of lexicographically sorted motifs found by the extraction tool. The i-th cell of the array A stores the list of pointers to motifs, lexicographically sorted, occurring at position i of input sequences and stored in the list L: these motifs share at least one occurrence, that is the one at position i. The array A can be built either during the extraction process of motifs with a constant additional cost, or in a post-processing stage. Thanks to the array A the proposed filter performs the intersections of occurrence sets for couples of motifs which share at least an occurrence.

After computing the intersection of occurrence sets of two motifs, if the size of the intersection is greater than the parameter ζ the filter checks the conditions of Definition 3 to determine which is the (ζ, δ)-conserved motif, and removes the other from the list of motifs.

Note that two motifs sharing several occurrences are present in lists stored at each position of the array A at which both motifs occur. The filter however can avoid to compute the intersection of occurrence sets for motifs already compared in the following manner: when it computes the intersection of occurrence sets of two motifs stored at position i of the array A, if the first occurrence shared by both motifs is at position $j < i$, then it means that the two motifs have already been compared.

As we already said, artifacts of a motif m are motifs sharing a subset of occurrences between themselves and with m, and that are less conserved than m. The filter proposed in this section to check the (ζ, δ)-conservation of motifs, compares at each iteration either an artifact with the real motif, or two artifacts of the same motif. In the second case, if one artifact is (ζ, δ)-conserved with respect to the other, only one artifact is kept. As we will see in Section 5 we can always choose values for thresholds ζ and δ such that at a successive iteration, the real motif is compared with previously kept artifacts and is (ζ, δ)-conserved with respect to them.

5 Experiments

In this section we validate the notion of maximality in conservation for approximate motifs proposed in this paper. In particular, we show results obtained applying the post-processing filter, described in Section 4, to the output provided by the motif extraction tool *SMILE* [9,8] in order to remove redundance due to the presence of artifacts of real motifs and hence to return maximally conserved motifs only.

We applied the filter on 5 fragments of the chromosome XXII of *Homo sapiens* composed of 480 bases. We chose these fragments because they are rich of repetitions, exact and approximate, in number significantly large with respect to random sequences, and very long (up to 35-50 base pairs). The Table 1 reports the number of motifs common to all the sequences in the input set found by *SMILE* admitting just one substitution between motifs and their occurrences and varying motif length from 4 to 22 base pairs. We show the running time of *SMILE* as a reference in evaluating the time taken by our post-processing step which will be showed in Section 5.2, and not to provide an evaluation of the performance of *SMILE*.

Table 1. Number of motifs and running time of the extraction tool *SMILE*

MOTIF LENGTH	QUORUM	ADMITTED SUBSTITUTIONS	N° FOUND MOTIFS	TIME
4	100	1	101	0.02 sec
8	100	1	459	0.04 sec
12	100	1	1216	0.08 sec
16	100	1	2222	0.11 sec
20	100	1	3279	0.11 sec
21-22	100	1	7196	0.24 sec

We applied our post-processing filter for the check of the maximality in conservation in order to remove artifacts from the motifs found by traditional tools for the extraction of approximate motifs. The notion of maximality in conservation of motifs given in Section 2 depends on two parameters ζ and δ: in next section we will discuss the behaviour of our filter as these parameters change.

5.1   Setting Filter Parameters

We saw in Section 4 that the goal of the post-processing filter is to identify and remove artifacts of real motifs in order to increase significancy and readability of motif extraction results. To this purpose, motifs sharing at least ζ occurrences are compared and the ones which are (ζ, δ)-conserved with respect to the others are kept while the others are removed from the output. In this section we will see how many and which motifs are removed by the filter depending on the parameters ζ and δ.

Parameter ζ. The parameter ζ represents the minimum number of occurrences that two motifs must share so that the (ζ, δ)-conservation of a motif with respect to the other makes sense. The best choice for this parameter clearly also depends on the input sequences and, in particular, on the number of occurrences of sought motifs. Note that in real biological sequences it is possible to find less frequent motifs sharing occurrences with more frequent motifs. In that case, having just a single parameter which determines the number of occurrences that two motifs must share in order to check the (ζ, δ)-conservation of a motif with respect to the other is not effective, because, for example, referring the parameter to the less frequent motif, there is the risk to compare also motifs sharing very few occurrences (if the difference between the number of occurrences of the two motifs is large). On the contrary, in order to identify artifacts of real motifs, it is more efficient to use two parameters: one referred to the number of occurrences of the less frequent motif, and the other referred to the number of occurrences of the more frequent motif. In our implementation of the proposed post-processing filter, these two parameters are defined as a percentage of the number of occurrences of the less frequent motif, and a percentage of the number of occurrences of the more frequent motif. As a consequence, these two parameters represent two lower bounds to the number of occurrences that two motifs must share in order to speak about (ζ, δ)-conservation of a motif with respect to the other. The filter compares two motifs only if the intersection of their occurrences is greater or equal to both bounds; in particular, two motifs are not compared if the number of occurrences of the less frequent motif is smaller than the bound referred to the more frequent motif (because in this case the intersection of occurrences has size certainly lower than this bound). It is obvious that the lower are these bounds, the more artifacts are removed by the filter, as shown in Tables 2 and 3. Observe that setting very high bounds, for example with the percentage of occurrences of the less frequent motif (called $MIN\ BOUND$) and the one of the most frequent motif (called $MAX\ BOUND$) fixed to 90% and 80%, respectively, the filter compares motifs nearly equally frequent and that share many occurrences. Therefore, the filter removes motifs which are surely artifacts based on the notion of (ζ, δ)-conservation.

Decreasing the bound to the number of occurrences that two motifs must share referred to the more frequent motif and maintaining high the one referred to the more frequent motif, the filter compares motifs having an increasing number of occurrences in common. In this way, the filter may remove also artifacts of a motif

Table 2. Number of removed motifs and running time of the post-processing filter, applied to 459 motifs of length 8, as the percentage of shared occurrences referred to the more frequent motif ($MAX\ BOUND$) changed, and fixed the percentage of shared occurrences referred to the less frequent motif ($MIN\ BOUND$) to 90% and $\delta = 0.050$

MAX BOUND	N° REMOVED MOTIFS	N° MAXIMALLY CONSERVED MOTIFS	TIME
80%	320	139	417 msec
70%	330	129	368 msec
65%	335	124	342 msec
60%	353	106	308 msec
55%	359	100	276 msec

which, in addition of being less conserved, are less frequent with respect to the real motif, but they still share a large number of occurrences with it (remember that in some cases the real motif is less frequent with respect to its artifacts and for this reason we introduced the parameter δ, as we will see in the next section). For example, for tests in which we applied the filter to motifs of length 8 found by $SMILE$, setting parameters $MIN\ BOUND$ and $MAX\ BOUND$ to high values (that is, 90% and 80%, respectively) and $\delta = 0.050$, all the artifacts of the motif $m = ATCACTGC$ are removed. Indeed, m has 77 occurrences and it is the most frequent and/or the most conserved with respect to all motifs with number of occurrences between 66 and 74 sharing all their occurrences with m (hence, the occurrence intersection between each pair of motifs compared by the filter has size in percentage of 100% with respect to the less frequent motif and more than 85% with respect to the more frequent motif), therefore these motifs are all artifacts of m. The motif m, however, does not represent the real functional element which is present in the input sequences, because the 96% (that is, 74) of its occurrences is shared with the motif $m' = ATCACTA$ which is more conserved than m ($\mu_m = 0.96$ and $\mu_{m'} = 0.41$, this indicates that m and all its artifacts are artifacts of m'). As m' has 113 occurrences, we have to decrease the percentage of the number of shared occurrences referred to the more frequent motif from 85% to 65% so that the filter identifies m' as (ζ, δ)-conserved with respect to m.

On the contrary, decreasing the bound to the number of occurrences that two motifs must share referred to the less frequent motif, the filter compares motifs having a decreasing number of occurrences in common, therefore a larger number of artifacts are removed by the filter. For example, for tests in which we applied the filter to motifs of length 8 found by $SMILE$, setting parameters $MIN\ BOUND$ and $MAX\ BOUND$ to high value (that is, 80% and 55%, respectively) and $\delta = 0.050$, some artifacts of the motif $m = ATCACTGC$ are removed, but artifacts $m' = CCACCACT$ and $m'' = TCACCACC$ are not identified, because the number of occurrences shared between these artifacts and m is less than 80%. If we decrease $MAX\ BOUND$ to 75%, then m is compared

Table 3. Number of removed motifs and running time of the post-processing filter, applied to 459 motifs of length 8, as the percentage of shared occurrences referred to the less frequent motif ($MIN\ BOUND$) changes, and fixed the percentage of shared occurrences referred to the more frequent motif ($MAX\ BOUND$) to 55% and $\delta = 0.050$

MIN BOUND	N° REMOVED MOTIFS	N° MAXIMALLY CONSERVED MOTIFS	TIME
90%	359	100	276 msec
80%	381	78	261 msec
75%	388	71	257 msec
65%	394	65	246 msec
60%	397	62	237 msec

also with m' and m'', which are both less frequent and less conserved than m. For example, for m and m'' which have 546 and 402 occurrences, respectively, and share 302 occurrences, we have $\mu_m = 0.49$ e $\mu_{m''} = 0.93$.

As we already observed in Section 2, decreasing too much both bounds may cause the filter to compare motifs which do not share many occurrences and hence they do not represent the same functional element. Therefore, there is the risk to remove real motifs and not only artifacts. The biologists should choose appropriate bounds based on the biological sequences to be analyzed and on the kind of functional element to be sought.

Parameter δ. As already observed, in some cases the functional element to be sought is represented by a motif more conserved but less frequent than its artifacts. The parameter δ provides a bound to the minimum difference between the conservation degree of two motifs. Similarly to the case of the parameter ζ, a too small value for δ could cause the removal of motifs which are not artifacts, because the filter might remove motifs which are less conserved than the motif with which they are compared, but this could just depend on the fact that they have more occurrences and not on a really smaller conservation.

Table 4. Number of motifs removed by the post-processing filter, applied to 459 motifs of length 8, $MAX\ BOUND = 55\%$ and $MIN\ BOUND = 60\%$ and as the parameter δ changes

δ	N° REMOVED MOTIFS	N° MAXIMALLY CONSERVED MOTIFS	TIME
0.050	397	62	237 msec
0.030	400	59	237 msec
0.026	402	57	237 msec
0.016	405	54	229 msec
0.010	408	51	227 msec

To clarify these aspects, we now consider some significant examples observed during our tests. Applying the post-processing filter to the set of motifs of length

8 found by $SMILE$ admitting just one substitution between motifs and their occurrences, with parameters $MAX\ BOUND = 55\%$, $MIN\ BOUND = 60\%$ and $\delta = 0.030$, the filter compares the two motifs $m = GCATCACC$, having 60 occurrences and $\mu_m = 0.983$, and $m' = TCATCACC$, with 85 occurrences and $\mu_{m'} = 0.988$. The motifs m and m' share 59 occurrences (98.3% with respect to m and 69.4% with respect to m') and the difference between the Hamming distance means is equal to 0.004. The motif m has just an exact occurrence among 60, and m' has just an exact occurrence among 85, therefore the difference between the means of Hamming distances does not depend on a greater conservation of the less frequent motif with respect to the more frequent one, but just from the different number of occurrences of the two motifs. As a consequence, in this case, fixing $\delta = 0.004$ leads to determine that m is (ζ, δ)-conserved with respect to m', but it does not make sense because even though the two motifs share several occurrences, none of them is actually more conserved than the other. Using the same parameters, the filter compares motifs $m = CATCACTG$ e $m' = CATCACTT$ which have 79 and 77 occurrences, respectively, and all the occurrences of m' are also occurrences of m. However, m', even though is less frequent than m', is more conserved than m', because the difference between the means of Hamming distances which is equal to 0.026 ($\mu_{m'} = 0.935$ e $\mu_m = 0.962$) depends on the fact that m' has 5 exact occurrences among 77 while m has 3 exact occurrences among 79. Therefore, in this case, decreasing δ to 0.026 allows the filter to identify and to remove the motif m, whose information is contained in m', because m is only a less accurate representation of the functional element represented by m', that is m is an artifact of m'.

5.2   Running Time of the Filter

In this section we evaluate the performance in time of the proposed filter as parameters ζ and δ change.

Tables 2, 3, 4 and 5 show that the lower are thresholds ζ and δ, the faster is the filter to remove artifacts of motifs. This follows from the fact that, when the filter identifies a motif which is (ζ, δ)-conserved with respect to another, it removes the latter from the set of motifs to be filtered. As, the lower are thresholds ζ and δ, the more artifacts are identified, and therefore more motifs are removed, the time taken by the filter decreases, because it has to compare less motifs.

The running time of the filter depends on the number of motifs output by the tools of extraction as well. In general, as the Table 6 shows, the more motifs are given in input to the filter, the larger is the number of artifacts to remove, and hence the time taken by the filter to reduce or remove the redundance increases.

Another factor which influences the running time of the filter is the number of occurrenes of motifs to be compared, because, as explained in Section 4, in order to check the (ζ, δ)-conservation of a motif with respect to another one, the filter computes the intersection of their occurrence sets. Therefore, it follows that, the smaller is the size of occurrence sets, the faster is the filter to process the input.

Table 5. Running time of the post-processing filter, applied to 459 motifs of length 8, as MAX $BOUND$ and MIN $BOUND$ change, and with $\delta = 0.010$

MIN BOUND	MAX BOUND	N° REMOVED MOTIFS	N° MAXIMALLY CONSERVED MOTIFS	TIME
90%	80%	346	113	386 msec
80%	70%	364	95	332 msec
75%	65%	381	78	263 msec
65%	60%	401	58	232 msec
60%	55%	408	51	227 msec

Table 6. Running time of the post-processing filter, applied to motifs of different length, with parameters MAX $BOUND = 55\%$, MIN $BOUND = 60\%$ and $\delta = 0.010$, as the number of motifs change

MOTIF LENGTH	N° FOUND MOTIFS	N° REMOVED MOTIFS	N° MAXIMALLY CONSERVED MOTIFS	TIME
4	101	76	25	119 msec
8	459	408	51	227 msec
12	1216	1123	93	322 msec
16	2222	2078	144	537 msec
20	3279	3075	204	507 msec
21-22	7196	6719	477	974 msec

Taking into consideration the time needed to infer motifs (see Table 1) and the large number of artifacts which are removed by the proposed post-processing filter (see Table 6), we can say that it is possible to obtain more readable output with no redundancy in conservation, even starting from very large set of motifs, in negligible time. For example, in sequences on which we performed tests, $SMILE$ finds 7196 motifs of length between 21 and 22 in 0.24 seconds and our filter removes 6719 artifacts in 974 milliseconds, obtaining hence an output of just 477 maximally conserved motifs (with respect to the 7196 returned by $SMILE$) in less than one second.

6   Conclusions

The output of existing tool for the extraction of motifs might be very large when we look for approximate motifs. The reason is a redundancy in the output also due to the presence of artifacts of motifs representing real functional elements in the sequences. The artifacts of a motif are patterns of the same length of the real motif which represent the functional element less accurately than the real motif. In order to remove such kind of redundancy from the output of motif extraction tools, we defined a notion of motif maximality in conservation and we designed a method as alternative solution to the ones existing in literature and based on measures of statistical significance to separate artifacts from real motifs. We validated our notion of maximality in conservation by developing a post-processing filter which implements the proposed method and applying it to the output of an existing tool for the extraction of motifs from real biological sequences in order to remove artifacts.

References

1. Blanchette, M., Sinha, S.: Separating real motifs from their artifacts. In: Proceedings of the 9th Int. Conf. on Intell. Syst. for Mol. Biol., ISMB, Copenhagen, Denmark, July 21-25, pp. 30–38 (2001), Supplement of Bioinformatics
2. Federico, M., Pisanti, N.: Suffix tree characterization of maximal motifs in biological sequences. In: BIRD, pp. 456–465 (2008)
3. Federico, M., Pisanti, N.: Suffix tree characterization of maximal motifs in biological sequences. Theor. Comput. Sci. 410(43), 4391–4401 (2009)
4. Grossi, R., Pietracaprina, A., Pisanti, N., Pucci, G., Upfal, E., Vandin, F.: Madmx: A strategy for maximal dense motif extraction. J. of Comput. Biol. 18(4), 535–545 (2011)
5. Haubler, M.: Motif discovery on promotor sequences. Master's thesis, Institut fur Informatik and IRISA/INRIA Rennes, Universitat Potsdam, Supervised by Dr. Torsten Schaub and Dr. Jacques Nicolas (2005)
6. Kolpakov, R., Kucherov, G.: Finding approximate repetitions under hamming distance. In: Meyer auf der Heide, F. (ed.) ESA 2001. LNCS, vol. 2161, pp. 170–181. Springer, Heidelberg (2001)
7. Kurtz, S., Ohlebusch, E., Schleiermacher, C., Stoye, J., Giegerich, R.: Computation and visualization of degenerate repeats in complete genomes. In: Proceedings of the 8th Int. Conf. on Intell. Syst. for Mol. Biol. (ISMB), pp. 228–238 (2000)
8. Marsan, L., Sagot, M.-F.: Algorithms for extracting structured motifs using a suffix tree with an application to promoter and regulatory site consensus identification. J. of Comput. Biol. 7(3-4), 345–362 (2000)
9. Marsan, L., Sagot, M.-F.: Extracting structured motifs using a suffix tree - algorithms and application to promoter consensus identification. In: RECOMB, pp. 210–219 (2000)
10. Parida, L., Rigoutsos, I., Floratos, A., Platt, D.E., Gao, Y.: Pattern discovery on character sets and real-valued data: linear bound on irredundant motifs and an efficient polynomial time algorithm. In: SODA, pp. 297–308 (2000)
11. Pisanti, N., Crochemore, M., Grossi, R., Sagot, M.-F.: Bases of motifs for generating repeated patterns with wild cards. IEEE/ACM Trans. Comput. Biology Bioinform. 2(1), 40–50 (2005)
12. Soldano, H., Viari, A., Champesme, M.: Searching for flexible repeated patterns using a non-transitive similarity relation. Pattern Recognition Letters 16, 243–246 (1995)

A Approach to Clinical Proteomics Data Quality Control and Import

Pierre Naubourg, Marinette Savonnet,
Éric Leclercq, and Kokou Yétongnon

University of Burgundy
Laboratory LE2I - UMR5158
9 Avenue Alain Savary
21000 Dijon, France
{pierre.naubourg,marinette.savonnet,
eric.leclercq,kokou.yetongnon}@u-bourgogne.fr
http://le2i.cnrs.fr

Abstract. Biomedical domain and proteomics in particular are faced with an increasing volume of data. The heterogeneity of data sources implies heterogeneity in the representation and in the content of data. Data may also be incorrect, implicate errors and can compromise the analysis of experiments results. Our approach aims to ensure the initial quality of data during import into an information system dedicated to proteomics. It is based on the joint use of models, which represent the system sources, and ontologies, which are use as mediators between them. The controls, we propose, ensure the validity of values, semantics and data consistency during import process.

Keywords: Data Quality, Import, Ontology, Information System.

1 Introduction

Our research framework is in the biomedical domain and more specifically in clinical proteomics. Generally, proteomic platforms study proteins of organisms. Specifically, clinical proteomics is the study of characteristics of proteins in samples collected from groups of patients participating in a clinical study. A typical example is the discovery of biomarkers: 1) to identify and classify diseases, 2) to make early detection or diagnosis, and 3) to measure the response of patients to a therapy. The workflow of proteomic platforms is based on proteomics studies, involving a large number of samples data, from which the proteomic platforms extract relevant characteristics through experiments. In addition to the data used by the proteomic experiments, e.g. data resulting from mass spectrometer analysis, the statistical studies carried out after these experiments require accessing large volume of clinical data ranging from patient's characteristics and samples to diagnosed pathologies, transport conditions, and the conditions of storage of samples.

Proteomic platforms commonly use *Laboratory Information Management System* (LIMS) to manage different aspects of proteomics studies, varying from

C. Böhm et al. (Eds.): ITBAM 2011, LNCS 6865, pp. 168–182, 2011.

storing clinical information to realizing statistical studies following experiments on various equipments. There is a direct link between the imported clinical data in the LIM and the conclusions of a proteomic study. So, increasing the quality of data during the import process will lead to increase the accuracy of analysis results.

The information in the LIMS can easily be polluted by missing, redundant or even incorrect data without an effective management of data quality. Researchers (industries and academics) are increasingly interested in data quality [6,21]. For many years, methods of prevention, audit and data cleaning are used to improve data quality in information systems. Berti-Équille lists four complementary approaches of data quality: diagnostic, adaptive, corrective and preventive ([3]). Diagnostic approaches mainly use statistical methods to detect errors in large amount of data. Adaptive approaches provide dynamic treatments for real time verification of constraints to ensure data quality. Corrective approaches attempt to detect errors by comparing data with real values and suggesting corrections. Finally, preventive approaches deal with evaluation of models and processes used in the LIMS.

To improve the quality of data in the context of proteomic information management, we propose a semi-automated data import method to garantee the quality of the data imported in the LIMS is not altered by changing the context of usage. To deal with these issues, our approach is based on three differents levels of controls. The first level deals with data source problems. These problems, often linked due to the particularities of the partner information systems, involve conflicts arise from the concepts manipulated by the systems. The next two levels are centered on data usage problems, which appear when data do not validate the context where they are imported. The context corresponds, on one hand, to data management within the LIMS, and on the other hand, to the domain logic. One level of control is used to ensure that data are complete and coherent as regards to the LIMS by checking constraints linked with the model. The other level of control checks data coherence according to the domain logic. This level deals with the creation of rules from the domain ontology to validate or unvalidate some data.

In the remainder of this paper, we illustrate through examples, in section 2, the issues of clinical data import. In section 3 we present tools and methods used in our approach. Sections 4 and 5 present our approach and its implementation in clinical proteomics. Finally, section 6 concludes this paper and offers opportunities we plan on this work.

2   Data Import Issues

This section presents the context in which we conduct our work and several issues related to data import. Data are provided by external collaborators to the platform (called "partner" in the rest of the paper). Partners can be, for example, clinicians who own pathological files, University Hospitals which store biological characteristics about patients, or Biobanks which organize the storage of samples, etc. Thus, for each proteomic study, the analyzed samples are associated to clinical data that must be imported from external sources into the LIMS.

To show the key characteristics of heterogeneity, we present some relevant examples of datasets received by a proteomic platforms. Tables 1 and 2 are extracted from clinical datasets provided to the proteomics platform by two clinicians (C1 and C2 respectively). Each row of the tables is associated with a biological sample. Data import issues can be divided in two categories: multiplicity of data sources and data usage.

Table 1. C1 clinician dataset (extract)

SampleNum	PatientNum	Sex	Birth	Pathology	Organ
S124	HG65	G	may-26-07	LAL	bone marrow
S125				LAL	bone marrow
S126	HG65	B	may-26-07	LAL	bone marrow
S127	PM37	B	juil-01-07	LAL	bone marrow

Table 2. C2 clinician dataset (extract)

SampleCode	BirthDay	PatientCode	Gender	Disease	Location
654	08/16/48	hj25	F	neoplasm of breast	breast
HG12	02/01/62	hu65	F	neoplasm of breast	breast
S7	04/12/56	JH34	M	neoplasm of breast	liver
YK37	02/29/45	dv12	F	neoplasm of breast	breast

2.1 Problems Related to the Multiplicity of Data Sources

Partners, providing datasets, work in differents ways on samples. These various views on samples imply heterogeneity in terms of the datasets they provide to the proteomics platform.

Semantic conflicts have been studied extensively by researchers [15,19,26]. In summary, Goh identified three types of semantic heterogeneity [10]: 1) naming conflicts that occur in the presence of homonyms and synonyms, 2) scaling conflicts that arise when description granularities are not the same, and 3) confusion conflicts that arise when a word is used with two different meanings by two actors. Degoulet, working on message exchanged among actors of the biomedical domain, has highlighted the possibility of solving problems of semantics through the use of controlled vocabularies [8].

Data Semantics

The datasets in table 1 and 2 show various vocabularies for columns names. Clinician C1 (Table 1) uses *Birth* while C2 clinician (Table 2) uses *Birthday*. Obviously, the semantics of these two fields is the namely *the patient's date of birth*.

Data Format
We can also notice some differences on data formats. For example, in the case of birthday, clinician C1 chooses the format `mmm-DD-YY` while clinician C2 chooses the format `MM/DD/YY`. To match these data, data must be converted to the corresponding data format.

Field Values and Scale
Incompatible values (from different domains) can be used for corresponding fields in the tables. For example, the patient attribute about gender (*Sex* for clinician C1 and *Gender* for clinician C2), has as domain values {G,B} for clinician C1 (he is mainly working with children) and {M,F} for clinician C2.

Problems of scale can be divided in two categories. Measurement problems occur for example when a volume is expressed in μl and another in ml. The other problem concerns the granularity. For example, the same stage of development of a tumor can be described by several fields detailing the different characteristics of evolution in one source or by a single field that combine all characteristics in another source.

During the import process, these problems can be solved if the format and field of values are known and if automatic conversion methods are available. However, the problems related to semantics are much more complex and require technical representation of domain knowledge.

2.2   Problems Related to the Use of Data

The management of biological and biomedical data raises many information design problems. Chen identified four technological challenges in the field of genomics [5]: 1) complexity of data (due to various granularities reflecting various aspects and specialities), 2) specialized knowledge (needed for capturing their semantics), 3) continuing evolution of knowledge and 4) variety of profiles of people (working in bioinformatics and trying to reach a consensus to meet a common goal). Among these challenges, data complexity is the most challenging problem in our context. Biological data are complex because they are heterogeneous [7], incomplete, uncertain and inconsistent [31]. Despite these characteristics, the expertise of proteomic experiments requires high data quality to make pertinent conclusions.

Data completeness and coherence are a key concerns for many researchers. Chapter 2 of Han's book [13] gives a summary of different solutions dealing with missing or incorrect values. One solution is to ignore the tuple or the object. Another solution is to manually fill data gaps and modifify incorrect data. Other intermediates solutions use a constant, an average value or a decision tree to determine the missing data.

Completeness
As illustrate in table 1, sample S125 (second row in table 1) exhibits missing values for the fields used to identify the corresponding patient. Two solutions can be considered: reject the data due to the lack of identification values or use an annotation to distinguish the invalid data from others.

Coherence

Data describing the same concept can sometimes define different characteristics for the concept. For example, in Table 1 even though samples S124 and S126 refer to the same patient (HG65), gender is Boy in S124 and Girl in S126, defining two differents genders for the same patient.

Domain Logic

The domain knowledge can highlight another problem. For example, the data in table 2 concern a proteomics study of breast cancer. Most samples of this datasets are taken from the breast of the patient. However, the sample S7 (third row in the table 2) is traken from the patient's liver. Is this a mistake made by the clinicain or a new detail that need to be studied ? Only a domain expert can answer this question. An implementation of a knowledge base to represent the rules defining the domain logic can be used to detect inconsistencies in the data.

To implement our approach, we need: 1) a model representing domain knowledge; 2) a model representing business knowledge (i.e. the business logic of the proteomics platform); 3) a model of the LIMS and 4) the schema of data sources.

3   Background

Linster presents two views on model building: 1) modelling to make sense and 2) modelling to implement systems [16]. Modeling to make sense is used to formally organize domain knowledge whereas modeling to implement, the most commonly used, consits in organizing the components of a system to execute them on a computer. In our proposal, we combine the two approaches by using ontologies to represent knowledge and models to implement data quality module.

3.1   Ontologies

According to [11], an ontology is an explicit specification of a conceptualization . In practice, ontologies can be used to represent domain knowledge or as an aid to understand a system by separating data and domain concepts. There are various types of ontologies used for specific purposes. Van Heijst, define four types of ontologies ([29]): generic, domain, representation and application. In this article, we will only discuss domain and application ontologies.

Domain Ontology

Domain ontology is used to represent consensual knowledge in a domain ([24]). It represents the key concepts of the domain linked by various relationships. The main relations used are specializations (*is-a*), synonyms and the generic relations (*related-to*). This type of ontology is used to ensure the consistency of semantics (also called *semantic net* by Wiederhold in [30]) among various systems. Domain ontology can serve as scientific reference in exchanges with partners. The concepts and relationships are then used as a syntactic and semantic consensus.

Many efforts are made in the biomedical field for structuring knowledge in the form of ontologies. The *Gene Ontology consortium*[1] produces a controlled vocabulary in the form of an ontology about roles of genes in protein expression ([1]).

Given the dynamic nature of knowledge, we chose to implement an evolving system to manage domain logic. Our system is based on "rules" defined on relationships among concepts of the domain ontology. Concerning information systems, business rules are formal expressions that constrain some aspects of a system. They structure, control and influence a system ([12,22]). Recent works have shown the benefits of rules for Semantic Web ([17,14]). In our approach we focus on rules for defining new part of knowledge that are not directly modeled in the ontology. Only domain experts can define pertinents rules to be taken into account to increase proteomics platform knowledge. The evolving characteristic of the rules system is given by decoupling knowledge (ontologies and rules) and implementation of the system.

Application Ontology

An application ontology is used to represent the knowledge of implemented systems. Compared to domain ontologies, application ontologies respresent the reality of the information systems to which they are affiliated. An ontology of this type can be used in a system of cooperation among various partners in a domain. It often serves as a reference for technical meetings among system users, to determine if a concept of a system corresponds to another concept of another system. For example, two systems with patient identifiers, `PatientNum` and `PatientCode`, will refer to the same concept `PatientId` of the application ontology. In our approach, this type of ontology is used as a mediator among partners and LIMS schema.

3.2   Models

Models are representations of systems according to certain points of view. Among the modeling languages, one of the most used is probably the Unified Modeling Language (UML). UML defines several diagrams to describe several aspect (structural, behavioral, temporal, etc.) of a system or an application. Fowler defines three ways to use UML models in his book *"UML Distilled"* [9]: as *sketches*, as *blueprints* or as a *programming language*. According to Fowler, UML models are used mainly as sketches to help the understanding of ideas among project participants during meetings. They aren't focused on development. Blueprints are precise enough to be implemented by a developer. Using UML as a programming language allows immediate implementation of UML models into executable code: diagrams become the program's source code. In our approach, UML models are defined as blueprints, they will be accurate enough to be implemented by simple transformation into executable code.

---

[1] `http://www.geneontology.org`

3.3 Coupling Ontologies and Models

Spear ([27]) defines two dimensions for the construction of a domain description:

– the horizontal dimension (or relevance) determines the scope of information that must be included in the representation of knowledge;
– the vertical dimension (or granularity) determines the accuracy of the representation of knowledge.

Ontologies, due to their mechanism of refinement and specialization are best suited to the vertical dimension of a domain. The horizontal axis is better supported by models that allow the aggregation of knowledge over large areas.

Ashenhurst asserts that the use of ontologies to guide semantics and thus the domain knowledge is relevant [2]. Our proposal incorporates these findings by using ontologies to support knowledge modeling and UML models (mainly class diagram) to define structure of system components.

4 Organization of Data Quality Components

Our approach is mainly based on the use of ontologies as mediators among partner systems and LIMS system. The controls made during data import can check and detect some errors following three steps. The first step is to check semantics, domain and data format using an application ontology. The second step is to verify data completeness and coherence through the use of the components structure defined in the UML class diagram. The last step is to check business rules related to the domain knowledge. Once these three steps are performed, the validated data can be stored in the LIMS database. Figure 1 represents a summary view of models and ontologies organization used during this process.

4.1 Clinical Data Model Used in the LIMS

The LIMS used by the proteomics platform maintains data in a relational database which can store *identified* and if necessary *transformed* data to ensure the relevance of search tools and data quality.

Clinical data model was realized by using UML class diagram and presents patient-specific data and their associations to pathologies (via a date of diagnosis, a patient may present several diseases) and to biological data samples. To store ontological information, we add domain "classifications" used by proteomic platforms. Diseases can be associated with a code complying to the International Classification of Diseases[2] proposed by the World Health Organisation. The class diagram follows the ICD structure *Chapter - Section - Element* to allow a more or less fine description. For example, a clinician may define a disease by ICD code C78.7 (Secondary malignant neoplasm of the liver) or by the code C00-D48 (malignant tumors) according to the accuracy of information provided . The cancer tumors may be associated with a code TNM (Tumor, Nodes, Metastasis) to define the extent of tumor in a patient's body.

---

[2] International Classification of Diseases (ICD),
   http://www.who.int/classifications/icd

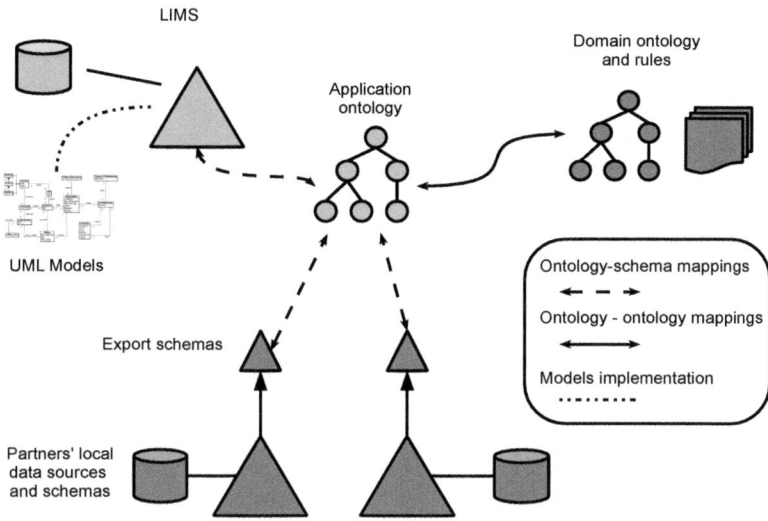

Fig. 1. Summary view of models, ontologies and mappings organization

4.2   Ontologies

Two ontologies are needed in our approach: a domain ontology to support the domain knowledge and an application ontology to support specific partners knowledge.

Domain Ontology

The construction of this ontology followed a method based on "relevant questions" and by searching common concepts in the domain. According to Brusa [4], relevant questions are questions posed by experts during their "investigations" and that the ontology can provide an answer for. Here is an example of a relevant question: " Can I know the extent of this tumor ?". The other aspect of the construction of this ontology is based on the finding of common concepts ([28]).

Figure 2 presents an extract from the domain ontology. The resource consensus that we have chosen to respond to relevant questions are CIM, TNM nomenclature, the branch of anatomy of MeSH and recommendations of the National Cancer Institute (INCA) in tumors banks[3]. This recommendation includes common concepts of clinical data.

The rules, we use in our approach, are based on associations among concepts of domain ontology. An example of "associations for rules" is shown on Figure 2. It specifies which organs are affected by diseases. For this, we define a generic relation *affectedOrgan* linking the concept *Anatomy* (from the MeSH branch) and the concept *Disease* (from the ICD branch). Then, the expert must "specialize"

---

[3] Tumour banks are banks of cryopreserved tumor tissues.

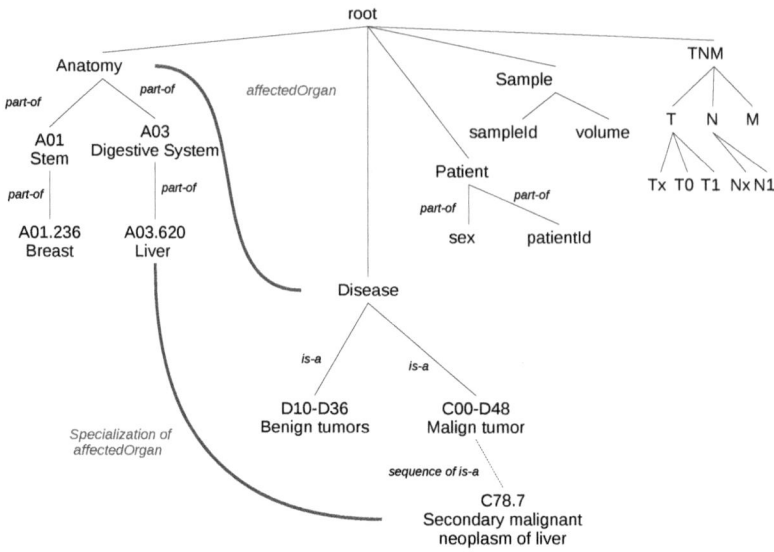

Fig. 2. Domain ontology (extract)

knowledge by defining which organs are affected by diseases: e.g. the *Liver* is an organ affected by the pathology *C78.7* (secondary malignant neoplasm of liver). A rule must then be created defining the validity of a sample if the pathology and the organ are mutually relevant.

Application Ontology
The application ontology is used as a mediator between the models of partners and the model of the LIMS. It is designed in agreement with key partners and the proteomic platform. Each partners' schema has a match between the descriptors of data (classes, attributes, headers, etc.) and a concept of the ontology. Figure 3 is an extract of our application ontology.

4.3   Mappings

We borrow the concept of *mapping* used in ontology alignment works ([25,23]) to represent correspondences among concepts of two ontologies and among the concepts of the application ontology and the schema descriptors. We use two types of mappings: ontological mappings between two concepts of ontologies and ontology-schema mappings linking a ontological concept to a schema descriptor.

Ontological Mappings
Ontological mappings M_O are mappings of type 1..1 to express an equivalence between concepts. In our approach, this mapping is used to match the concepts of the application ontology to those of the domain ontology. The mappings

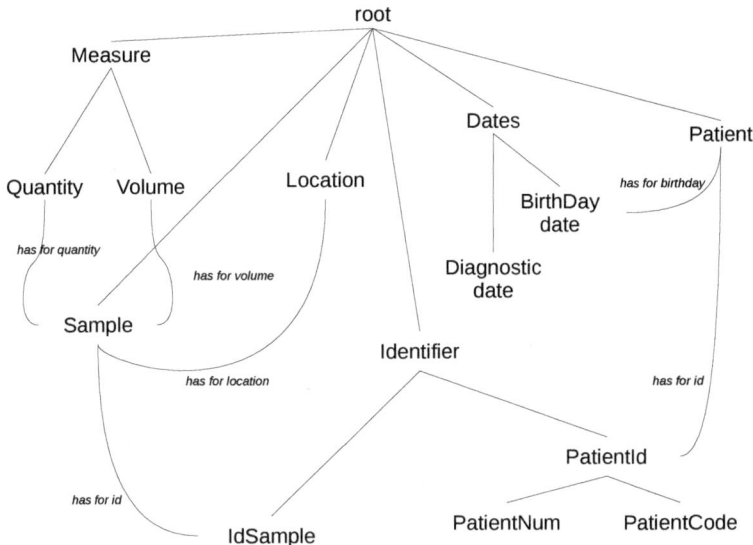

Fig. 3. Application ontology (extract)

are made during the construction of two ontologies and must be updated when one (or both) ontology (ies) evolve. For example, we have created the following ontological mapping: M_O1 ($Anatomy_{DO}$, $Location_{AO}$) to match the concept *Anatomy* of the domain ontology DO and the concept *Location* of application ontology AO.

Definition 1. *An ontological mapping M_O is a pair $\langle C_{o1}, C'_{o2} \rangle$ where C is a concept of an ontology o1 and C' is a concept of an ontology o2.*

We decide to make a loose coupling among application ontology and domain ontology because of their different degree of evolution. The domain ontology is not set to change often, because its concepts are adopted by many experts. The application ontology can be extended and modified at each arrival (possibly departure) of a partner. The loose coupling among these two ontologies allows us, when modifying an ontology, to not impact the other concepts.

Ontology-Schema Mappings

Ontology-schema mappings M_{OS} link the concepts of an application ontology to data schemas descriptors. The mappings can be of type 1..1 linking one concept of an ontology to one descriptor of the schema, type 1..n linking one concept of an ontology to several descriptors of the schema, or type n. .1 linking several concepts from ontology to one single descriptor. The mappings define what is the exact meaning of each schema descriptor.

Definition 2. *A ontology-schema mapping M_{OS} is a pair $\langle \{D_S\}, \{C_o\} \rangle$ composed of a set of descriptors D from the schema S and a set of C concepts of ontology o.*

For example, the below are two ontology-schema mappings:

- $M_{OS}1$ ($NumPatient_{LIMS}, PatientId_{AO}$) which allows to link the *NumPatient* from the LIMS schema and the concept *PatientId* of the application ontology AO;
- $M_{SO}2$ ($\{Tumor_{P1}, Node_{P1}, Meta_{P1}\}, TNMStage_{AO}$) which allows to link the three descriptors *Tumor*, *Node* and *Meta* form the P1 partner's schema and the concept *TNMStage* of the application ontology AO.

Descriptors of schemas are also linked by ontology-schema mappings with the data formats branch of the application ontology. For example in our LIMS, the descriptor *BirthDate* is mapped to the format DD/MM/YYYY while the birthday date of the schema of partner 1 (*Birth*) is linked to the format DD-MM-YY. So we have two types of ontology-schema mappings: 1) to define the meaning of the descriptors and 2) to define the data format. The joint use of these both types of mappings allows to find the conversion function required to transform values.

Each schema has its specific characteristics. The entry of a new partner in this system may in some cases be made without changing the application ontology. We only have to perform ontology-schema mappings among descriptors and application ontology. In other cases, it is necessary to change the application ontology concepts impacted by specializing concepts. Ontology-schema mappings corresponding to other partners will not be impacted by such changes. For example, if a new partner is defining the location of samples by the use of two descriptors, we can expand the concept of *Location* of application ontology in two "sub-concepts": *Position* and *Depth*.

5   Implementation of the Approach

The implementation of our approach has three main steps. The first step involves the creation of objects based on the semantic definition and format of the data. The second step is to check coherence and completeness of the objects in accordance with the schema of our LIMS. The third and final step is to check the consistency of objects according to the domain logic. Figure 4 summarizes the various steps of our approach, for reasons of clarity, we do not show mappings present in Figure 1.

The first control concerns the semantics and data format. It uses ontology-schema mappings to determine semantics of each descriptor. Comparison of mappings performed on the LIMS' schema to those made on the partners' schema, hilights: 1) the correspondences among partners and LIMS descriptors, and 2) the conversion operations required to transform data values. The construction of objects is based on these two pieces of information. At the end of this step, we have "syntax objects".

Once the objects are created, we can check coherence and completeness. The use of UML class diagram as a structural model of our system allows you to specify optional and mandatory associations between objects. Thus we can identify association errors between objects. We can also verify the consistency of some

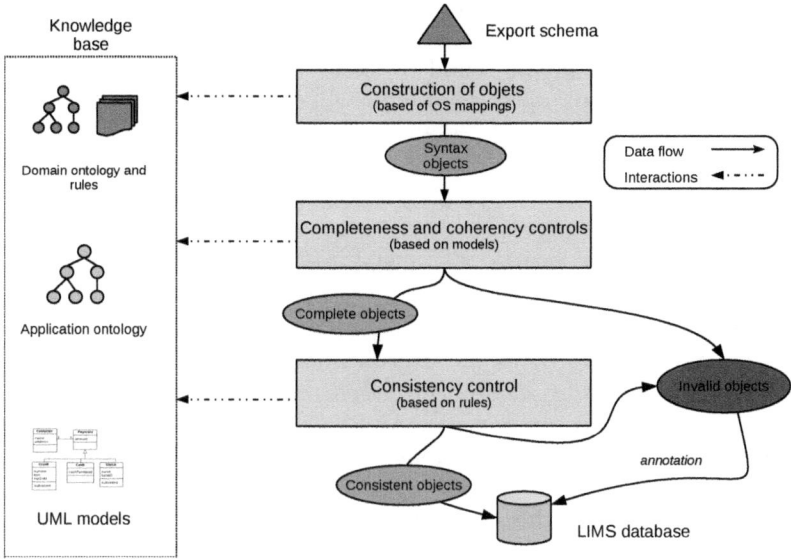

Fig. 4. Data flow in our approach

data within objects. Biological material is rare, we can not reject all of the invalid data. Invalid objects are inserted into the database with an annotation. For example, the clinician at the source of data set will be questioned to determine the gender of the patient. The annotation prevents the use of the biological sample within an experiment.

Once the objects checked, the rule engine takes into account the facts, i.e. the newly created objects and knowledge, and rules. At the end of this process we obtain consistent objects that have successfully passed three controls, or we obtain invalid objects. The rules supported by our implementation of the engine are written in SWRL ([14]) in accordance with the DL-Safe restriction [18]. For example, the following rule: *"a sample is valid if the disease for which it is studied and if the organ from which it comes are mutually relevant"* will be defined as:

```
Sample(?s), affectedOrgan(?o,?d), disease(?d)
   => ValidSample(?s)
```

The implementation of our approach describes in this article is included in the Clinical Module eClims[4] of open source LIMS ePims[TM]. Due to the confidentiality characteristic of proteomics data, we only could test our processes on only one dataset provided to the CLIPP[5] platform by a clinician. This dataset

---

[4] Further information and screenshots are available on the website: **http://eclims.u-bourgogne.fr**

[5] CLIPP: CLinical and Innovation Proteomic Platform. **http://www.clipproteomic.fr**

is a CSV file containing 345 samples and 64 relevant descriptors. We identified 114 samples which do not match overall quality. 9 of these 114 samples were not consistent and the rule engine found problems concerning the sex of patients. The remaining 105 samples present some problems of completeness.

6 Conclusion

Our data import system ensures the initial quality of clinical proteomics data. The implementation may require a major human investment especially during the ontologies creation. But this initial investment guarantee to each dataset coming from one source, the same overall quality. As our approach is center on the LIMS' system, the scalability of this method is acceptable because of the centralization of the components. Adding new sources, "only" require the creation of new ontology-schema mappings between the source schema and the application ontology.

The main perspective is the automatic creation of ontology-schema mappings, especially during the addition of a new partner. This improvement would almost allow complete automation of our approach. To this end, we are interested in papers related to automatic alignment of ontologies ([20]).

Acknowledgments. The authors wish to thank the proteomics platform CLIPP, the Company ASA (Advanced Solutions Accelerator) and the Regional Council of Burgundy for their supports.

References

1. Ashburner, M., Ball, C.A., Blake, J.A., Botstein, D., Butler, H., Cherry, J.M., Davis, A.P., Dolinski, K., Dwight, S.S., Eppig, J.T., Harris, M.A., Hill, D.P., Issel-Tarver, L., Kasarskis, A., Lewis, S., Matese, J.C., Richardson, J.E., Ringwald, M., Rubin, G.M., Sherlock, G.: Gene ontology: tool for the unification of biology. the gene ontology consortium. Nature genetics 25(1), 25–29 (2000)
2. Ashenhurst, R.L.: Ontological aspects of information modeling. Minds and Machines 6, 287–394 (1996)
3. Berti-Équille, L.: Quality Awareness for Data Managing and Mining. Habilitation à diriger les recherches, Université de Rennes 1, France (June 2007)
4. Brusa, G., Caliusco, M. L., Chiotti, O.: A process for building a domain ontology: an experience in developing a government budgetary ontology. In: Proceedings of the Second Australasian Workshop on Advances in Ontologies AOW 2006, Darlinghurst, Australia, Australia, vol. 72, pp. 7–15. Australian Computer Society, Inc. (2006)
5. Chen, J.Y., Carlis, J.V.: Genomic data modeling. Inf. Syst. 28, 287–310 (2003)
6. Dasu, T., Johnson, T.: Exploratory Data Mining and Data Cleaning. John Wiley, Chichester (2003)
7. Davidson, S., Overton, C., Buneman, P.: Challenges in Integrating Biological Data Sources. Journal of Computational Biology 2(4), 557–572 (1995)

8. Degoulet, P., Fieschi, M., Attali, C.: Les enjeux de l'interopérabilité sémantique dans les systèmes d'information de santé. Informatique et gestion médicalisée 9, 203–212 (1997)
9. Fowler, M.: UML Distilled: A Brief Guide to the Standard Object Modeling Language, 3rd edn. Addison-Wesley Longman Publishing Co., Inc., Boston (2003)
10. Goh, C.H.: Representing and reasoning about semantic conflicts in heterogeneous information systems. PhD thesis (1997)
11. Gruber, T.R.: Toward principles for the design of ontologies used for knowledge sharing. International Journal of Human-Computer Studies 43(5-6), 907–928 (1995)
12. Hall, J., Healy, K., Ross, R.: Defining Business Rules: What Are They Really? Rapport (2000)
13. Han, J., Kamber, M.: Data mining: concepts and techniques, 2nd edn. Morgan Kaufmann, San Francisco (2006)
14. Horrocks, I., Patel-Schneider, P.F.: A proposal for an owl rules language. In: Proceedings of the 13th International World Wide Web Conference (WWW 2004), pp. 723–731. ACM Press, New York (2004)
15. Kim, W., Seo, J.: Classifying schematic and data heterogeneity in multidatabase systems. Computer 24, 12–18 (1991)
16. Linster, M.: Viewing knowledge engineering as a symbiosis of modeling to make sense and modeling to implement systems. In: Ohlbach, H.J. (ed.) GWAI 1992. LNCS, vol. 671, pp. 87–99. Springer, Heidelberg (1993)
17. Motik, B., Rosati, R.: Reconciling description logics and rules. J. ACM 57, 1–30 (2008)
18. Motik, B., Sattler, U., Studer, R.: Query Answering for OWL DL with rules. Web Semantics 3(1), 41–60 (2005)
19. Naiman, C.F., Ouksel, A.M.: A classification of semantic conflicts in heterogeneous database systems. J. Organ. Comput. 5, 167–193 (1995)
20. Rahm, E., Bernstein, P.A.: A survey of approaches to automatic schema matching. The VLDB Journal 10, 334–350 (2001)
21. Redman, T.C.: Data quality: the field guide. Digital Press, Newton (2001)
22. Ross, R.G.: Principles of the Business Rule Approach. Addison-Wesley Longman Publishing Co., Inc., Boston (2003)
23. Safar, B., Reynaud, C., Calvier, F.-E.: Techniques d'alignement d'ontologies basées sur la structure d'une ressource complémentaire. In: 1ères Journées Francophones sur les Ontologies (JFO 2007), pp. 21–35 (2007)
24. Salem, S., AbdelRahman, S.: A multiple-domain ontology builder. In: Proceedings of the 23rd International Conference on Computational Linguistics, COLING 2010, Stroudsburg, PA, USA, pp. 967–975. Association for Computational Linguistics (2010)
25. Shvaiko, P.: Ten challenges for ontology matching. In: Chung, S. (ed.) OTM 2008, Part II. LNCS, vol. 5332, pp. 1164–1182. Springer, Heidelberg (2008)
26. Siegel, M., Madnick, S.E.: A metadata approach to resolving semantic conflicts. In: Proceedings of the 17th International Conference on Very Large Data Bases, VLDB 1991, pp. 133–145. Morgan Kaufmann Publishers Inc, San Francisco (1991)
27. Spear, A.D.: Ontology for the twenty first century: An introduction with recommendations. Institute for Formal Ontology and Medical Information Science, Saarbrücken, Germany (2006)
28. Sugumaran, V., Storey, V.C.: Ontologies for conceptual modeling: their creation, use, and management. Data Knowl. Eng. 42, 251–271 (2002)

29. Van Heijst, G., Schreiber, A.T., Wielinga, B.J.: Using explicit ontologies in KBS development. Int. J. Hum.-Comput. Stud. 46, 183–292 (1997)
30. Wiederhold, G.: Interoperation, mediation, and ontologies. In: Proceedings International Symposium on Fifth Generation Computer Systems (FGCS94), Workshop on Heterogeneous Cooperative Knowledge-Bases, vol. 3, pp. 33–48 (1994)
31. Willson, S.J.: Measuring inconsistency in phylogenetic trees. J. Theor. Biol. 190, 15–36 (1998)

MAIS-TB: An Integrated Web Tool for Molecular Epidemiology Analysis

Patricia Soares, Carlos Penha Gonçalves, Gabriela Gomes, and José Pereira-Leal

Instituto Gulbenkian de Ciência, Rua da Quinta Grande 6,
Apartado 14, P-2781-901
Oeiras, Portugal
jleal@igc.gulbenkian.pt

Abstract. There is growing evidence that the genetic diversity of *Mycobacterium tuberculosis* may have important clinical consequences. In that sense combining genetic, clinical and demographic data will allow a comprehensive view of the epidemiology of bacterial pathogens and their evolution, helping to explain how virulence and other phenotypic traits evolve in bacterial species over time. [1-2] Hence to understand TB, an integrative approach is needed.

Therefore we created MAIS-TB (Molecular Analysis Information System - TB), an informatics system, which integrates molecular analysis of MTB isolates, with clinical and demographic information. This system provides a new tool to access and identify associations between tuberculosis strain types and clinical and epidemiological characteristics of the disease.

Keywords: Tuberculosis, framework, molecular epidemiology.

1   Introduction

Tuberculosis is the second highest cause of death from an infectious disease worldwide and it is estimated that one third of the world's population is infected, although the majority will never develop active disease. The emergence of MDR (multi drug resistance) and XDR (extremely drug resistance) is threatening to make TB incurable [3-4].

A variety of social and biologic factors foster the accelerated progression and transmission of tuberculosis. But if and how *Mycobacterium tuberculosis* genomic diversity influences human disease in clinical settings remain open questions. To understand the complexity of the interactions between host, pathogen and environment, an integrative system is needed. [1] Some databases on tuberculosis and/or infectious diseases are available, although none of them provide the complete clinical and demographic picture, not allowing the understanding of the molecular mechanisms leading from strain genotype to clinical phenotypes.

We created MAIS-TB (Molecular Analysis Information System – TB), linking molecular, clinical and demographic data on portuguese patients. The isolates were genotyped for each one of the standard methods: SNP, MIRU, RFLP and Spoligotype.

C. Böhm et al. (Eds.): ITBAM 2011, LNCS 6865, pp. 183–185, 2011.
© Springer-Verlag Berlin Heidelberg 2011

2   Implementation

The data is stored in a MySQL database and the web interface is based on the Django web framework, running on Apache with mod_WSGI. Its written in Python and the phylogenetic trees are built with Biopython and NetworkX. For full functionality, Javascript needs to be enabled. This system is compatible with the common browsers and its possible to install in Linux, Windows and Mac.

We designed this system in a modular way, in which different types of data and different functionalities are stored and implemented separately. This modular architecture make it simple to adapt to other diseases (figure 1).

Each patient could have one or more infections (episodes), and one or more samples. On the other hand a sample only belongs to one patient. All the information, clinical and demographic, are connected through the episode number, and it's possible to obtain information about resistance or molecular data through the episode or the sample.

To preserve the security of the clinical and demographic data several access were established. Some users may access all the functionality while others can't download clinical data or enter new data into the system.

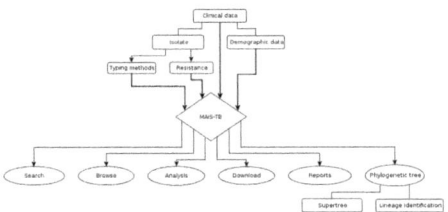

Fig. 1. Database schema

3   Results and Discussion

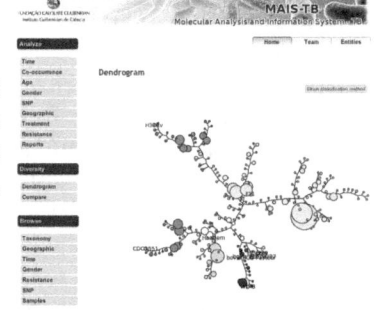

The system was designed to be used both in research labs (e.g. our own) as well as by health authorities (e.g. our collaborators). With this in mind we developed a system with many features, from which we highlight:

i)   Multiple ways to insert data, all with validation in every field: through a pipeline, uploading a file or filling a web-form.

Fig. 2. Screenshot of the dendrogram

ii) Analysis tools (plots, dendrogram) with automatic updating (figure 2).
iii) Automated strain classification.
iv) Possibility to download the data used to create the graphics allowing researchers to do   more analysis using other tools.
v) Possibility to generate pre-defined reports.

Combining molecular data with epidemiological information will allow to identify strains of bacteria and investigate the determinants and distribution of disease. Together they can establish transmission links, identify risk factors for transmission, and provide an insight into the pathogenesis of tuberculosis.

References

1. Coscolla, M., Gagneux, S.: Does M. tuberculosis genomic diversity explain disease diversity? ScienceDirect (2010)
2. Thwaites, G., Caws, M., Chau, T., D'Sa, A., Lan, N., et al.: Relationship between Mycobacterium tuberculosis genotype and the clinical phenotype of pulmonary and meningeal tuberculosis. Journal of Clinical Microbiology (2008)
3. Millet, J., Badoolal, S., Akpaka, P., Ramoutar, D., Rastogi, N.: Phylogeographical and molecular characterization of an emerging M. tuberculosis clone in T & T. ScienceDirect (2009)
4. Caws, M., Thwaites, G., Dunstan, S., Hawn, T., Lan, N., et al.: The influence of Host and Bacterial Genotype on the Development of Disseminated Disease with Mycobacterium tuberculosis. PloS Pathogens (2008)

Author Index

GPSR Compliance

The European Union's (EU) General Product Safety Regulation (GPSR)
is a set of rules that requires consumer products to be safe and our
obligations to ensure this.

If you have any concerns about our products, you can contact us on
ProductSafety@springernature.com

In case Publisher is established outside the EU, the EU authorized
representative is:

Springer Nature Customer Service Center GmbH
Europaplatz 3
69115 Heidelberg, Germany

Batch number: 09474016

Printed by Printforce, the Netherlands